Walter Hood Fitch

Illustrations of the British Flora

A series of wood engravings, with dissections, of British plants

Walter Hood Fitch

Illustrations of the British Flora
A series of wood engravings, with dissections, of British plants

ISBN/EAN: 9783337256463

Printed in Europe, USA, Canada, Australia, Japan

Cover: Foto ©Andreas Hilbeck / pixelio.de

More available books at **www.hansebooks.com**

ILLUSTRATIONS

OF

THE BRITISH FLORA:

A SERIES OF

WOOD ENGRAVINGS, WITH DISSECTIONS,

OF

British Plants,

DRAWN BY

W. H. FITCH, F.L.S.

AND

W. G. SMITH, F.L.S.,

FORMING AN ILLUSTRATED COMPANION TO MR. BENTHAM'S HANDBOOK AND OTHER BRITISH FLORAS.

LONDON:
L. REEVE & Co., 5, HENRIETTA STREET, COVENT GARDEN.
1880.

PREFACE.

THE Illustrated Edition of Mr. Bentham's "Handbook of the British Flora" being exhausted, the Wood engravings of that work are here reproduced as an Illustrated Companion to the "Handbook" and other British Floras. The Cuts are arranged according to the last Edition of the "Handbook," and new Cuts of the species admitted in recent Editions are added. To facilitate reference from other Floras, where the nomenclature differs from that of the "Handbook," synonyms, in italics, are incorporated in the Index. In this volume and the "Handbook" combined, Students will have, in a more convenient and portable form, all that the Illustrated Edition contained, at little more than one-third the cost.

1. Clematis vitalba.

2. Thalictrum alpinum.

3. Thalictrum minus.

4. Thalictrum flavum.

5. Anemone pulsatilla.

6. Anemone nemorosa.

7. Adonis autumnalis.

8. Myosurus minimus.

9. Ranunculus aquatilis.

10. Ranunculus hederaceus.

11. Ranunculus lingua.

12. Ranunculus flammula.

13. Ranunculus ophioglossifolius.

16. Ranunculus auricomus.

14. Ranunculus ficaria.

17. Ranunculus acris.

15. Ranunculus sceleratus.

18. Ranunculus repens.

19. Ranunculus chærophyllus.

20. Ranunculus bulbosus.

21. Ranunculus philonotis.

23. Ranunculus arvensis.

22. Ranunculus parviflorus.

24. Caltha palustris.

25. Trollius europæus.

26. Helleborus viridis.

27. Helleborus fœtidus.

28. Aquilegia vulgaris.

29. Delphinium ajacis.

30. Aconitum napellus.

31. Actæa spicata.

32. Pæonia officinalis.

33. Berberis vulgaris.

34. Nymphæa alba.

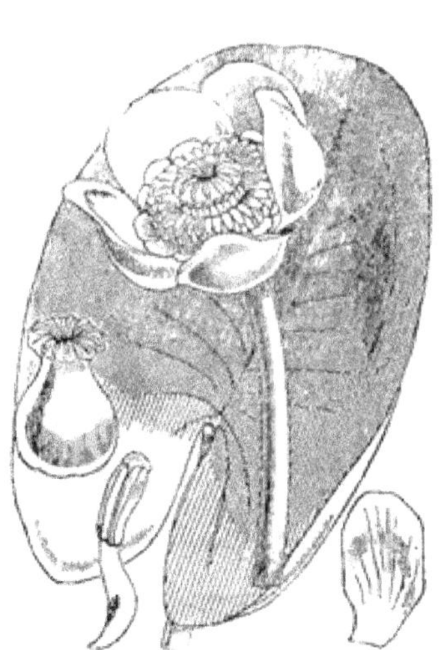

35. Nuphar lutea.

36. Papaver somniferum.

37. Papaver rhœas.

39. Papaver hybridum.

38. Papaver dubium.

40. Papaver argemone.

41. Meconopsis cambrica.

43. Rœmeria hybrida.

42. Chelidonium majus.

44. Glaucium luteum.

45. Fumaria officinalis.

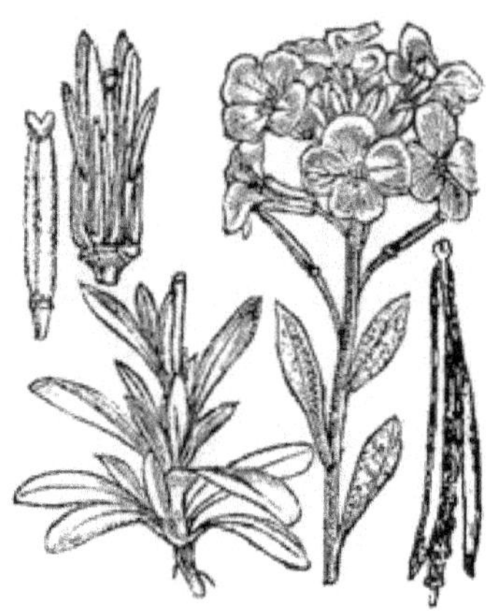

48. Matthiola incana.

46 Corydalis lutea.

47. Corydalis claviculata.

49. Matthiola sinuata.

50. Cheiranthus cheiri.

52. Nasturtium officinale.

51. Barbaria vulgaris.

53. Nasturtium sylvestre.

54. Nasturtium palustre.

55. Nasturtium amphibium.

56. Arabis perfoliata.

57. Arabis turrita.

58. Arabis hirsuta.

59. Arabis ciliata.

60. Arabis thaliana.

62. Arabis petræa.

61. Arabis stricta.

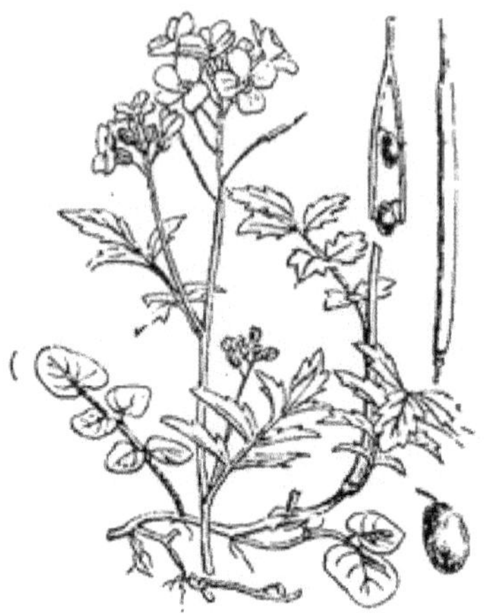

63. Cardamine amara.

64. Cardamine pratensis.

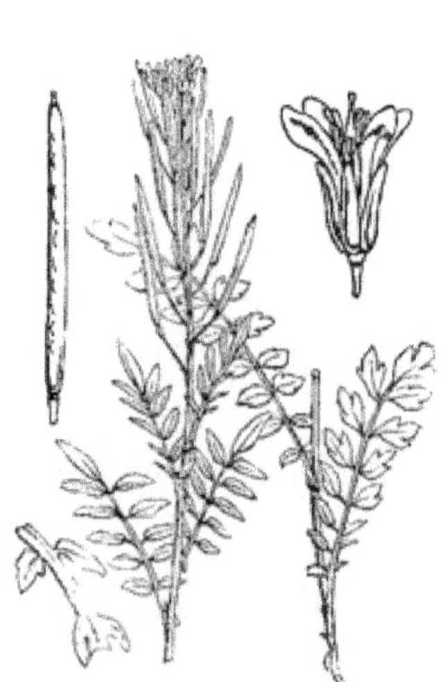

65. Cardamine impatiens.

66. Cardamine hirsuta.

67. Cardamine bulbifera.

68. Hesperis matronalis.

69. Sisymbrium officinale.

70. Sisymbrium irio.

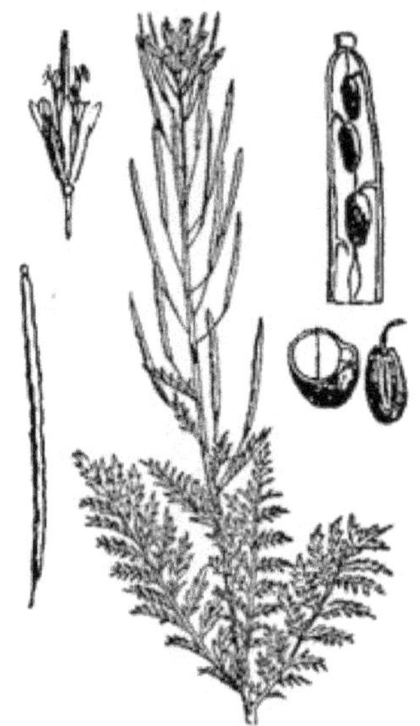

71. Sisymbrium sophia.

72. Alliaria officinalis.

73. Erysimum cheiranthoides.

74. Erysimum orientale.

75. Brassica tenuifolia.

77. Brassica monensis.

76. Brassica muralis.

78. Brassica oleracea.

79. Brassica campestris.

80. Brassica alba.

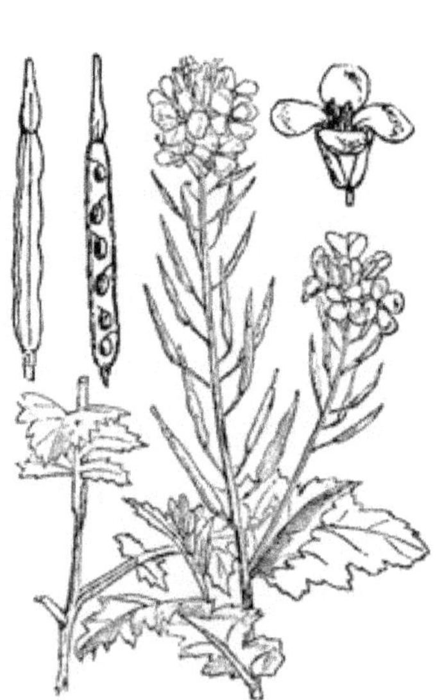

81. Brassica sinapistrum.

82. Brassica nigra.

83. Brassica adpressa.

84. Cochlearia armoracia.

85. Cochlearis officinalis.

86. Alyssum calycinum.

87. Alyssum maritimum.

90. Draba incana.

88. Draba aizoides.

91. Draba muralis.

89. Draba hirta.

92. Draba verna.

93. Camelina sativa.

94. Subularia aquatica.

96. Thlaspi perfoliatum.

95. Thlaspi arvense.

97. Thlaspi alpestre.

98. Teesdalia nudicaulis.

99. Iberis amara.

101. Capsella bursa-pastoris.

100. Hutchinsia petræa.

102. Lepidium campestre.

104. Lepidium draba.

103. Lepidium smithii.

105. Lepidium latifolium.

106. Lepidium ruderale.

108. Senebiera didyma.

107. Senebiera coronopus.

109. Isatis tinctoria.

110. Cakile maritima.

112. Raphanus raphanistrum.

111. Crambe maritima.

113. Reseda luteola

114. Reseda lutea.

115. Reseda alba.

116. Heliauthemum guttatum.

117. Helianthemum canum.

118. Helianthemum vulgare.

121. Viola odorata.

119. Helianthemum polifolium.

122. Viola hirta.

120. Viola palustris.

123. Viola canina.

124. Viola tricolor.

125. Polygala vulgaris.

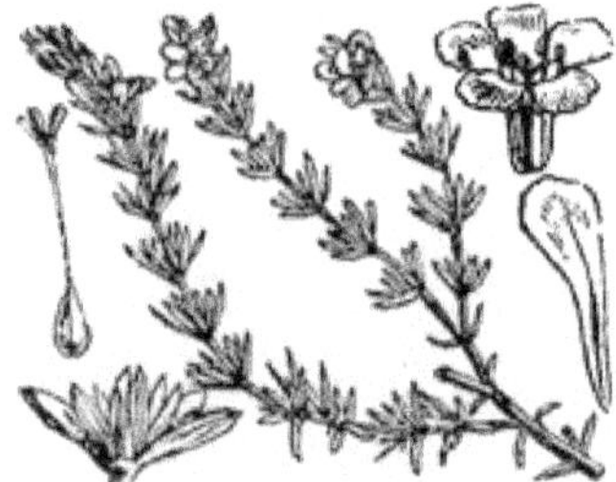

126. Frankenia lævis.

127. Dianthus prolifer.

128. Dianthus armeria.

129. Dianthus deltoides.

131. Saponaria officinalis.

130. Dianthus cæsins.

132. Silene acaulis.

133. Silene inflata.

134. Silene otites.

136. Silene gallica.

135. Silene nutans.

137. Silene conica.

138. Silene noctiflora.

139. Lychnis vespertina.

140. Lychnis diurna.

141. Lychnis githago.

142. Lichnis flos-cuculi.

143. Lychnis viscaria.

144. Lychnis alpina.

145. Sagina procumbens.

146. Sagina Linnæi.

147. Sagina nodosa.

148. Cherleria sedoides.

149. Arenaria verna.

150. Arenaria uliginosa.

152. Arenaria peploides.

151. Arenaria tenuifolia.

153. Arenaria serpyllifolia.

154. Arenaria ciliata.

155. Arenaria trinervis.

157. Holosteum umbellatum.

156. Mœnchia erecta.

158. Cerastium vulgatum.

159. Cerastium arvense.

161. Cerastium trigynum.

160. Cerastium alpinum.

162. Stellaria aquatica.

164. Stellaria media.

163. Stellaria nemorum.

165. Stellaria uliginosa.

166. Stellaria graminea.

167. Stellaria glauca.

168. Stellaria holostea.

169. Spergularia rubra.

170. Spergula arvensis.

171. Polycarpon tetraphyllum.

173. Montia fontana.

172. Claytonia perfoliata.

174. Tamaria gallica.

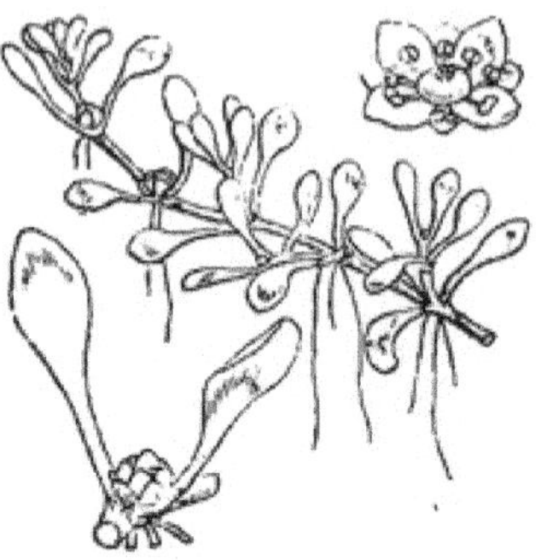

176. Elatine hydropiper.

175. Elatine hexandra.

177. Hypericum calycinum.

178. Hypericum androsæmum.

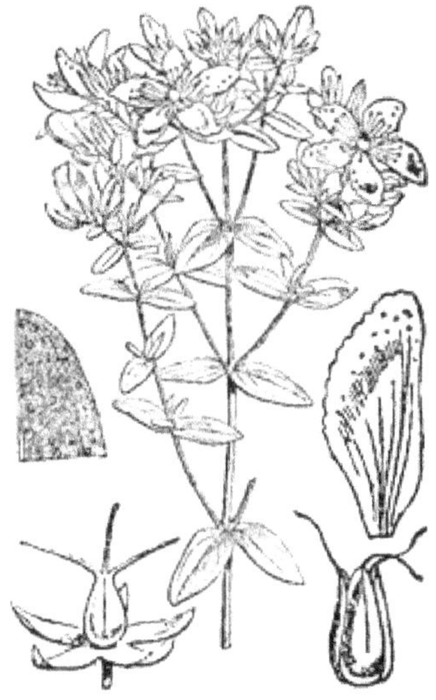

179. Hypericum perforatum.

180. Hypericum dubium.

181. Hypericum quadrangulum.

182. Hypericum humifusum.

183. Hypericum linariifolium.

184. Hypericum pulchrum.

185. Hypericum hirsutum.

186. Hypericum montanum.

188. Linum usitatissimum.

187. Hypericum elodes.

189. Linum perenne.

190. Linum angustifolium.

191. Linum catharticum.

192. Radiola millegrana.

193. Lavatera arborea.

194. Malva rotundifolia.

195. Malva sylvestris.

196. Malva moschata.

197. Althæa officinalis.

198. Althæa hirsuta.

199. Tilia europæa.

200. Geranium sanguineum.

201. Geranium phæum.

202. Geranium sylvaticum.

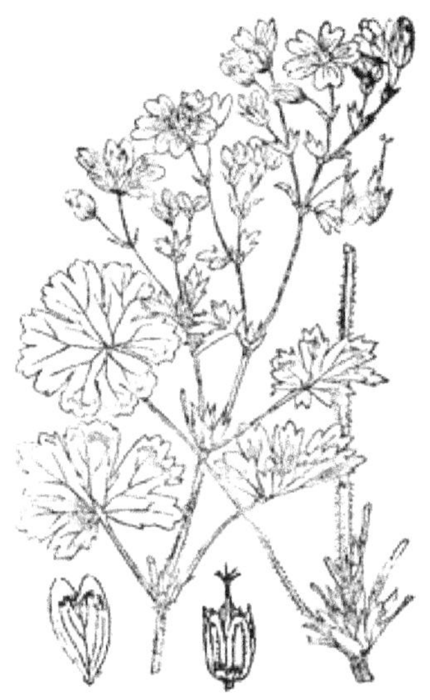

204. Geranium pyrenaicum.

203. Geranium pratense.

205. Geranium robertianum.

206. Geranium lucidum.

207. Geranium molle.

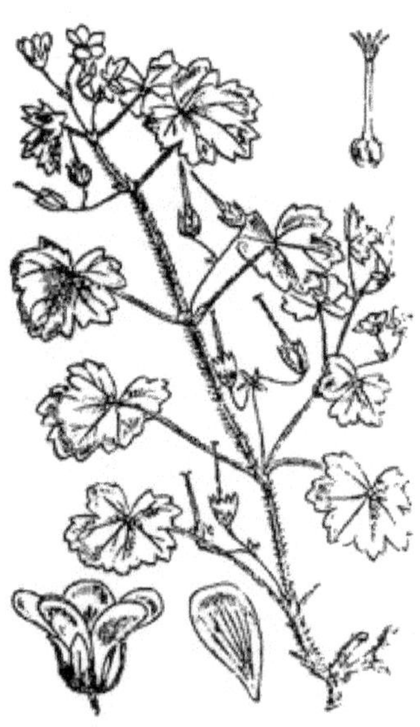

209. Geranium rotundifolium.

208. Geranium pusillum.

210. Geranium dissectum.

211. Geranium columbinum.

212. Erodium cicutarium.

213. Erodium moschatum.

214. Erodium maritimum.

215. Oxalis acetosella.

216. Oxalis corninulata.

217. Impatiens noli-me-tangere.

218. Impatiens fulva.

220. Acer pseudo-platanus.

219. Acer campestre.

221. Ilex aquifolium.

222 Evonymus europæus.

224. Rhamnus frangula.

223. Rhamnus calharticus.

225. Ulex europæus.

226. Ulex nanus.

228. Genista pilosa.

227. Genista tinctoria.

229. Genista anglica.

230. Cytisus scoparius.

231. Ononis arveusis.

232. Ononis reclinata.

233. Medicago falcata.

234. Medicago sativa.

235. Medicago lupulina.

236. Medicago denticulata.

237. Medicago maculata.

238. Medicago minima.

240. Melilotus arvensis.

239. Melilotus officinalis.

241. Melilotus alba.

242. Trigonella ornithopodioides.

244. Trifolium arvense.

243. Trifolium incarnatum.

245. Trifolium stellatum.

246. Trifolium ochroleucum.

247. Trifolium pratense.

248. Trifolium medium.

249. Trifolium maritimum.

250. Trifolium striatum.

252. Trifolium scabrum.

251. Trifolium bocconi.

253. Trifolium strictum.

254. Trifolium glomeratum.

255. Trifolium suffocatum.

256. Trifolium resupinatum.

258. Trifolium fragiferum.

257. Trifolium subterraneum.

259. Trifolium repens.

260. Trifolium hybridum.

261. Trifolium procumbens.

262. Trifolium minus.

263. Trifolium filiforme.

264. Lotus corniculatus.

265. Lotus angustissimus.

266. Anthyllis vulneraria.

267. Astragalus hypoglottis.

268. Astragalus alpinus.

269. Astragalus glycyphyllos.

270. Oxytropis campestris.

271. Oxytropis uralensis.

272. Ornithopus ebracteatus.

274. Hippocrepis comosa.

273. Ornithopus perpusillus.

275. Onobrychis sativa.

276. Vicia hirsuta.

277. Vicia tetrasperma.

278. Vicia cracca.

279. Vicia sylvatica.

280. Vicia Orobus.

281. Vicia sepium.

282. Vicia lutea.

283. Vicia sativa.

284. Vicia lathyroides.

285. Vicia bithynica

286. Lathyrus Nissolia.

287. Lathyrus Aphaca.

288. Lathyrus hirsutus.

290. Lathyrus tuberosus.

289. Lathyrus pratensis.

291. Lathyrus sylvestris.

292. Lathyrus palustris.

293. Lathyrus maritimus.

295. Lathyrus niger.

294. Lathyrus macrorrhizus.

296. Prunus communnis.

297. Prunus cerasus.

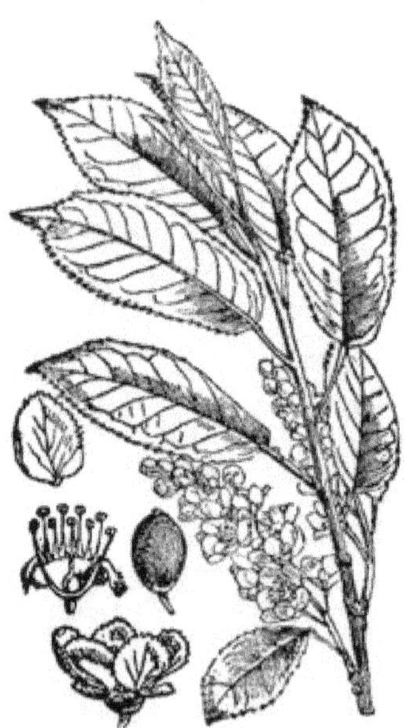

298. Prunus Padus.

299. Spiræa salicifolia.

300. Spiræa ulmaria.

301. Spiræa filipendula.

302. Dryas octopetala.

303. Geum urbanum.

304. Geum rivale.

305. Rubus idæus.

306. Rubus fruticosus.

307. Rubus cæsius.

308. Rubus saxatilis.

309. Rubus Chamæmorus.

311. Potentilla Fragariastrum.

310. Fragaria vesca.

312. Potentilla reptans.

313. Potentilla Tormentilla.

314. Potentilla argentea.

315 Potentilla verna.

316. Potentilla fruticosa.

317. Potentilla anserina.

318. Potentilla rupestris.

319. Potentilla comarum.

320. Sibbaldia procumbens.

321. Alchemilla vulgaris.

322. Alchemilla alpina.

323. Alchemilla arvensis

324 Sanguisorba officinalis.

325. Poterium Sanguisorba.

326. Agrimonia Eupatoria.

327 Rosa pimpinellifolia.

328. Rosa villosa.

329. Rosa rubiginosa.

330. Rosa canina.

331. Rosa arvensis.

332. Pyrus communis.

334. Pyrus Aria.

333. Pyrus Malus.

335. Pyrus torminalis.

336. Pyrus aucuparia.

338. Cotoneaster vulgaris.

337. Cratægus Oxyacantha.

339. Mespilus germanica.

340. Epilobium angustifolium.

341. Epilobium hirsutum.

342. Epilobium parviflorum.

343. Epilobium montanum.

344. Epilobium roseum.

346. Epilobium palustre.

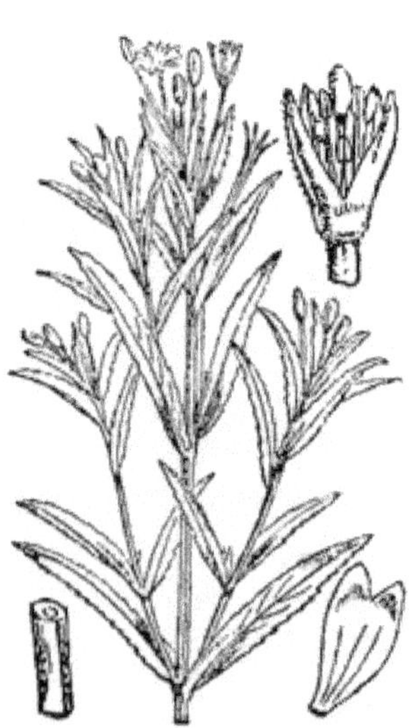

345. Epilobium tetragonum.

347. Epilobium alsinæfolium.

348. Epilobium alpinum.

350. Ludwigia palustris.

349. Œnothera biennis.

351. Circæa lutetiana.

352. Circæa alpina.

353. Lythrum Salicaria.

354. Lythrum hyssopifolium.

355. Peplis Portula

356. Bryonia dioica.

357. Tillæa muscosa.

358. Cotyledon umbilicus.

359. Sedum Rhodiola.

360. Sedum telephium.

361. Sedum anglicum.

362. Sedum dasyphyllum.

363. Sedum album.

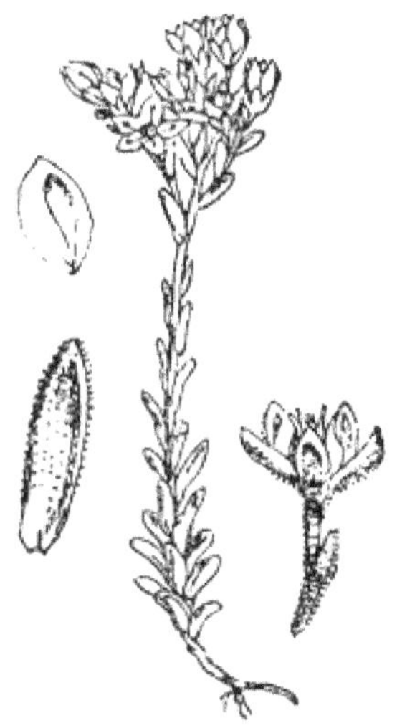

364. Sedum villosum.

365. Sedum acre.

366. Sedum sexangulare.

367. Sedum rupestre.

368. Sempervivum tectorum.

369. Ribes Grossularia.

370. Ribes rubrum.

371. Ribes alpinum.

372. Ribes nigrum.

374. Saxifraga aizoides.

373. Saxifraga oppositifolia.

375. Saxifraga Hirculus.

376. Saxifraga hypnoides.

378. Saxifraga granulata.

377. Saxifraga cæspitosa.

379. Saxifraga cernua.

380. Saxifraga rivularis.

381. Saxifraga tridactylites.

382. Saxifraga nivalis.

383. Saxifraga stellaris.

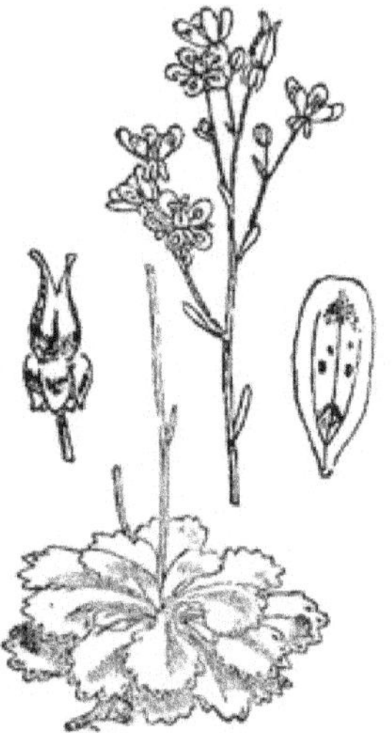

384. Saxifraga umbrosa.

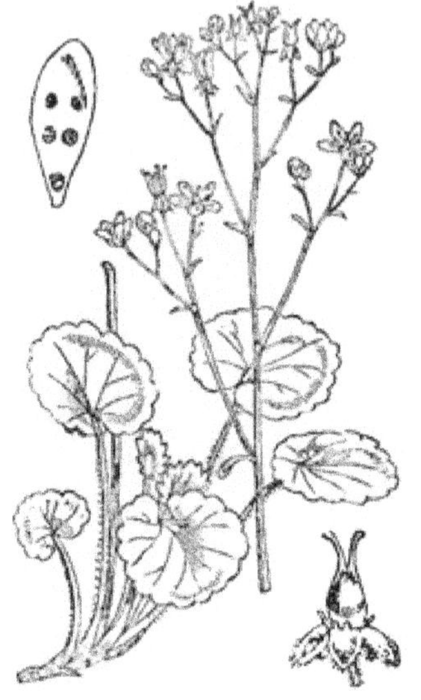

385. Saxifraga Geum.

386. Chrysosplenium oppositifolium.

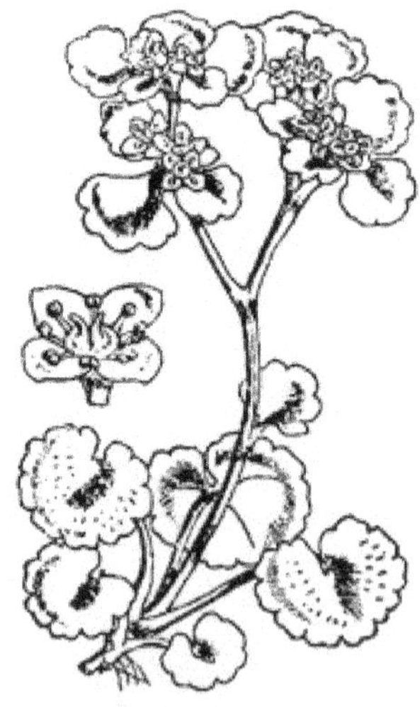

387. Chrysosplenium alternifolium.

388. Parnassia palustris.

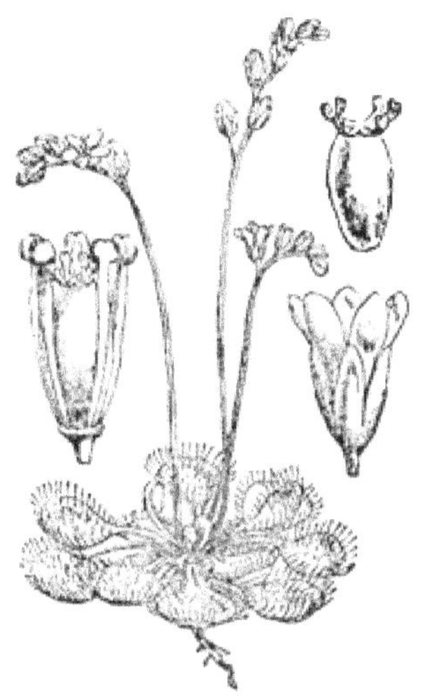

389. Drosera rotundifolia.

390. Drosera longifolia.

391. Drosera anglica.

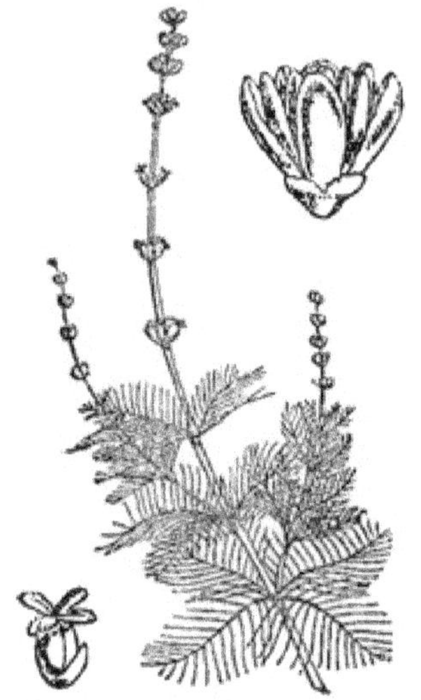

392. Myriophyllum spicatum.

393. Myriophyllum verticillatum.

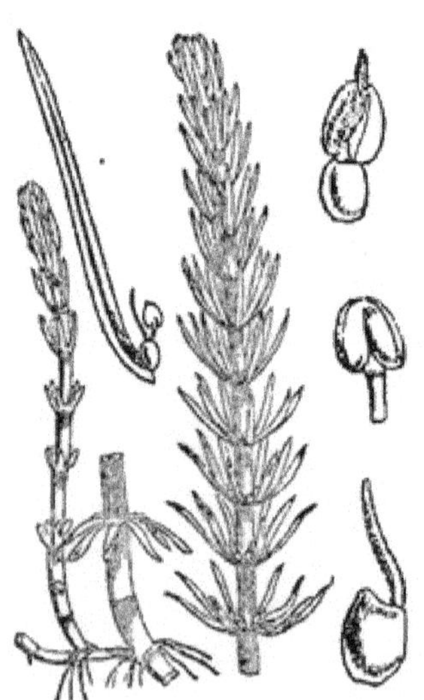

394. Hippuris vulgaris.

395. Hydrocotyle vulgaris.

396. Sanicula europea.

398. Eryngium maritimum.

397. Astrantia major.

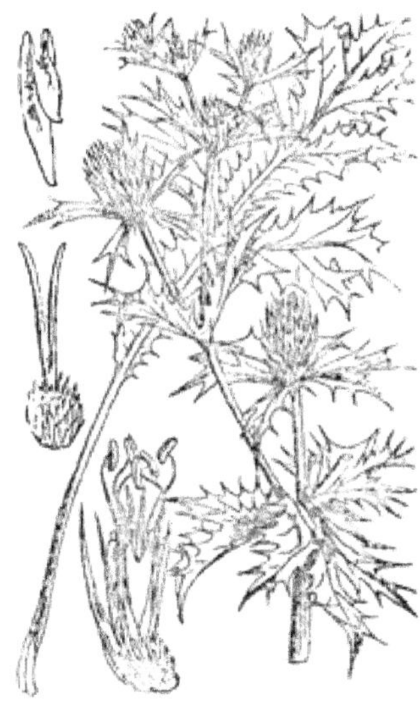

399. Eryngium campestre.

400. Cicuta virosa.

401. Apium graveolens.

402. Apium nodiflorum.

403. Apium inundatum.

404. Sison Amomum.

405. Trinia vulgaris.

406. Ægopodium Podagraria

407. Carum Petroselinum.

408. Carum segetum.

409. Carum verticillatum.

410. Carum Carvi.

411. Carum Bulbocastanum.

412. Sium latifolium.

413. Sium angustifolium.

414. Pimpinella Saxifraga.

415. Pimpinella magna.

416. Bupleurum rotundifolium.

417. Bupleurum aristatum.

418. Bupleurum tenuissimum.

419. Bupleurum falcatum.

420. Œnanthe fistulosa.

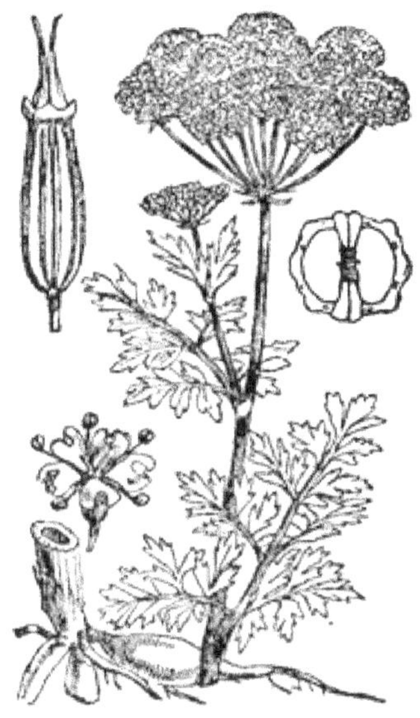

422. Œnanthe crocata.

421. Œnanthe pimpinelloides.

423. Œnanthe Phellandrium.

424. Æthusa Cynapium.

426. Seseli Libanotis.

425. Fœniculum vulgare.

427. Ligusticum scoticum.

428. Silaus pratensis.

429. Meum Athamanticum.

430. Chrithmum maritimum.

431. Angelica sylvestris.

432. Peucedanum officinale.

434. Peucedanum Ostruthium.

433. Peucedanum palustre.

435. Pastinaca sativa.

436. Heracleum Sphondylium.

438. Scandix Pecten.

437. Tordylium maximum.

439. Myrrhis odorata.

440. Conopodium denudatum.

442. Chærophyllum sylvestre.

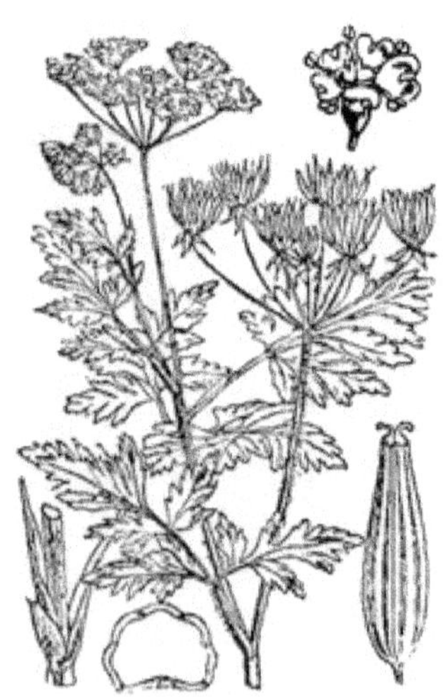

441. Chærophyllum temulum.

443. Chærophyllum Anthriscus.

444. Caucalis nodosa.

446. Caucalis infesta.

445. Caucalis Anthriscus.

447. Caucalis daucoides.

448. Caucalis latifolia.

449. Daucus Carota.

450. Conium maculatum.

451. Physospermum cornubiense.

452. Smyrnium Olusatrum.

453. Coriandrum sativum.

454. Hedera Helix.

455. Viscum album.

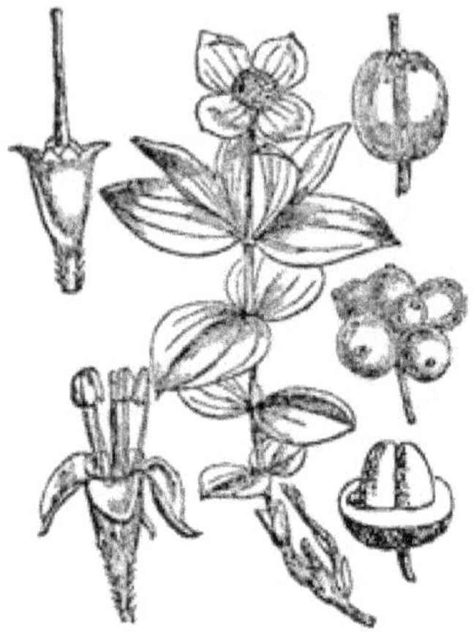

456. Cornus suecica.

457. Cornus sanguinea.

458. Adoxa Moschatellina.

459. Sambucus nigra

460. Sambucus Ebulus.

461. Viburnum Lantana.

462. Viburnum Opulus.

463. Lonicera Periclymenum.

464. Lonicera Caprifolium.

465. Lonicera Xylosteum.

466. Linnæa borealis.

467. Rubia peregrina.

468. Galium Cruciata.

469. Galium verum.

470. Galium palustre.

471. Galium uliginosum.

472. Galium saxatile.

473. Galium Mollugo.

475. Galium boreale.

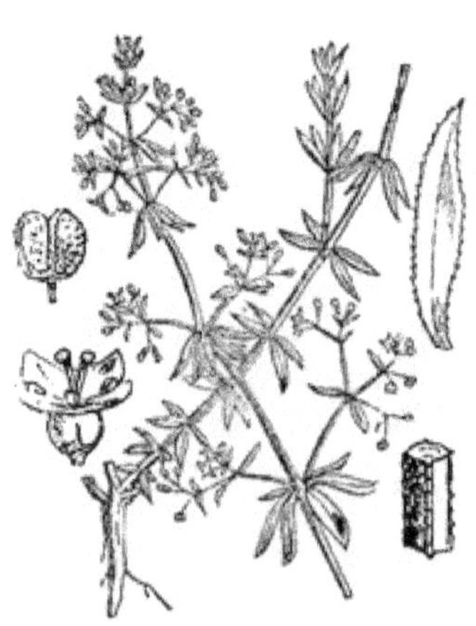

474. Galium parisiense.

476. Galium Aparine.

477. Galium tricorne.

478. Asperula odorata.

479. Asperula cynanchica.

480. Sherardia arvensis.

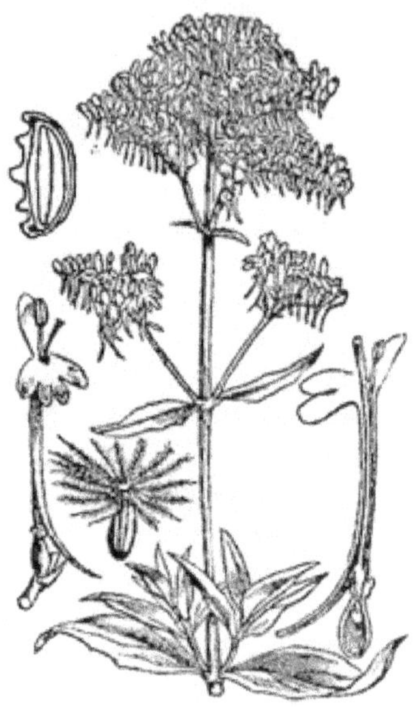

481. Centranthus ruber.

483. Valeriana officinalis.

482. Valeriana dioica.

484. Valeriana pyrenaica.

486. Valerianella carinata.

485. Valerianella olitoria.

487. Valerianella auricula.

488. Valerianella dentata.

489. Dipsacus sylvestris.

490. Dipsacus pilosus.

491. Scabiosa succisa.

492 Scabiosa Columbaria.

493. Scabiosa arvensis.

494. Eupatorium cannabinum.

495. Aster Tripolium.

496. Aster linosyris.

498. Erigeron alpinus.

497. Erigeron acris.

499. Erigeron canadensis.

500. Solidaga Virga-aurea.

501. Bellis perennis.

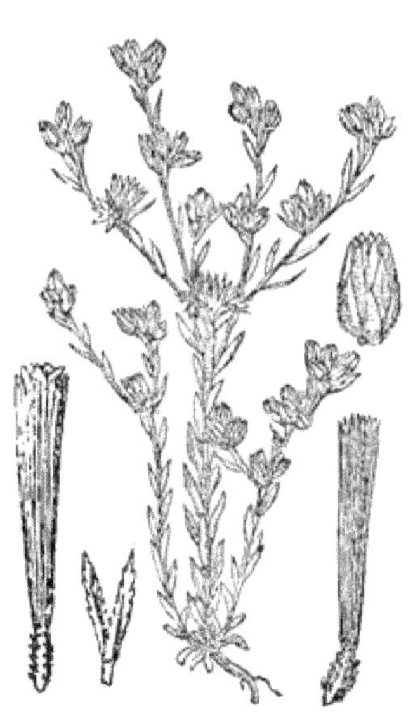

503. Filago minima.

502. Filago germanica.

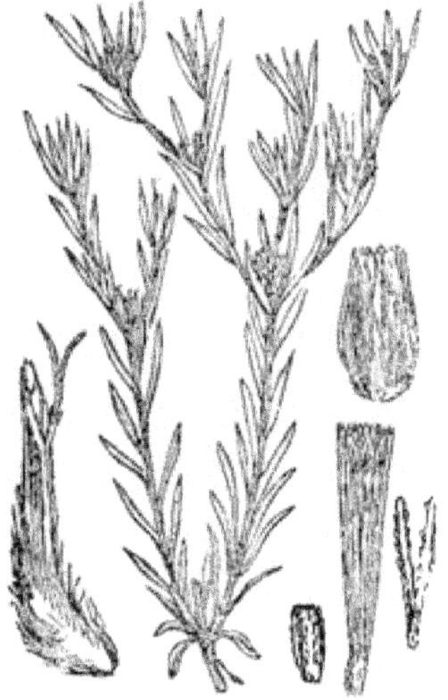

504. Filago gallica.

506. Gnaphalium sylvaticum.

505. Gnaphalium luteo-album.

507. Gnaphalium supinum.

508. Gnaphalium uliginosum.

509. Antennaria dioica.

510. Antennaria margaritacea.

511. Inula Helenium.

512. Inula salicina.

513. Inula crithmoides.

514. Inula conyza.

515. Inula dysenterica.

516. Inula Pulicaria.

517. Xanthium Strumarium.

518. Bidens cernua.

519. Bidens tripartita.

520. Chrysanthemum Leucanthemum

521. Chrysanthemum segetum.

522. Chrysanthemum Parthenium.

523. Matricaria inodora.

524. Matricaria Chamomilla.

526. Anthemis arvensis.

525. Anthemis Cotula.

527. Anthemis nobilis.

528. Anthemis tinctoria.

529. Achillea Ptarmica.

530. Achillea millefolium.

531. Diotis maritima.

532. Tanacetum vulgare.

533. Artemisia campestris.

534. Artemisia maritima.

535. Artemisia vulgaris.

536. Artemisia absinthium.

537 Tussilago farfara.

538. Tussilago petasites.

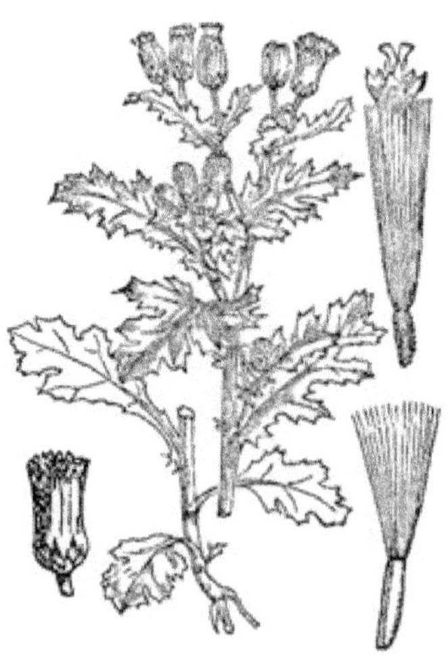

539. Senicio vulgaris.

540. Senecio viscosus.

541. Senecio sylvaticus.

542. Senecio squalidus.

543. Senecio aquaticus.

544. Senecio jacobæa.

545. Senecio erucifolius.

546. Senecio paludosus.

547. Senecio saracenicus.

548. Senecio palustris.

549. Senecio campestris.

550. Doronicum pardalianches.

551. Doronicum plantagineum.

552. Arctium lappa.

553. Serratula tinctoria.

554. Saussurea alpina.

555. Carduus marianus.

556. Carduus nutans.

557. Carduus acanthoides.

558. Carduus pycnocephalus.

559. Carduus lanceolatus.

560. Carduus palustris.

561. Carduus arvensis.

562. Carduus eriophorus.

563. Carduus heterophyllus.

564. Carduus tuberosus.

565. Carduus pratensis.

566. Carduus acaulis.

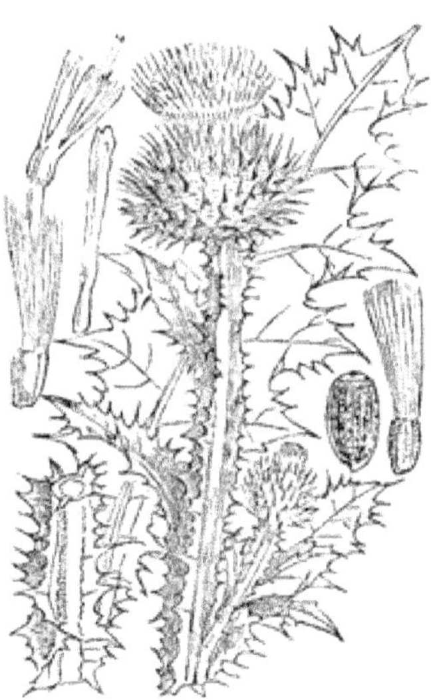

567. Onopordon acanthium.

568. Carlina vulgaris.

569. Centaurea nigra.

570. Centaurea scabiosa.

571. Centaurea cyanus.

572. Centaurea aspera.

573. Centaurea calcitrapa.

574. Centaurea solstitialis.

575. Tragopogon pratensis.

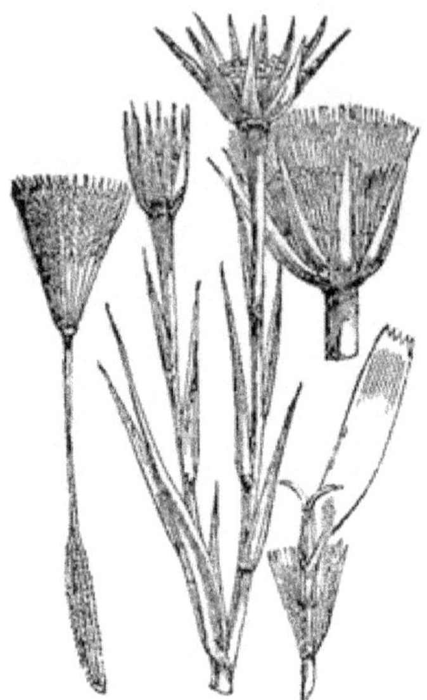

576. Tragopogon porrifolius.

577. Helminthia echioides.

578. Picris hieracioides.

579. Leontodon hispidus.

580. Leontodon autumnalis.

581. Leontodon hirtus.

582. Hypochœris glabra.

583. Hypochœris radicata.

584. Hypochœris maculata.

585. Lactuca muralis.

586. Lactuca scariola.

587. Lactuca saligna.

588. Sonchus arvensis.

589. Sonchus palustris.

590. Sonchus oleraceus.

591. Sonchus alpinus.

592. Taraxacum dens-levnis.

593. Crepis taraxacifolia.

594. Crepis fœtida.

595. Crepis virens.

596. Crepis biennis.

598. Crepis paludosa.

597. Crepis hieracioides.

599. Hieracium Pilosella.

600. Hieracium alpinum.

601. Hieracium murorum.

602. Hieracium cerinthoides.

603. Hieracium umbellatum.

604. Hieracium sabaudum.

606. Cichorium Intybus.

605. Hieracium prenanthoides.

607. Arnoseris pusilla.

608. Lapsana communis.

610. Lobelia urens.

609. Lobelia Dortmanna.

611. Jasione montana.

612. Phyteuma orbiculare.

614. Campanula glomerata.

613. Phyteuma spicatum.

615. Campanula Trachelium.

616. Campanula latifolia.

617. Campanula rapunculoides.

618. Campauula Rapunculus.

619. Campanula patula.

620. Campanula rotundifolia.

622. Campanula hybrida.

621. Campanula hederacea.

623. Vaccinium Myrtillus.

624. Vaccinium uliginosum.

625. Vaccinium Vitis-idæa.

626. Vaccinium Oxycoccos.

627. Arbutus unedo.

628. Arctostaphylos Uva-ursi.

630. Andromeda polifolia.

629. Arctostaphylos alpina.

631. Loiseleuria procumbens.

632. Menziesia polifolia.

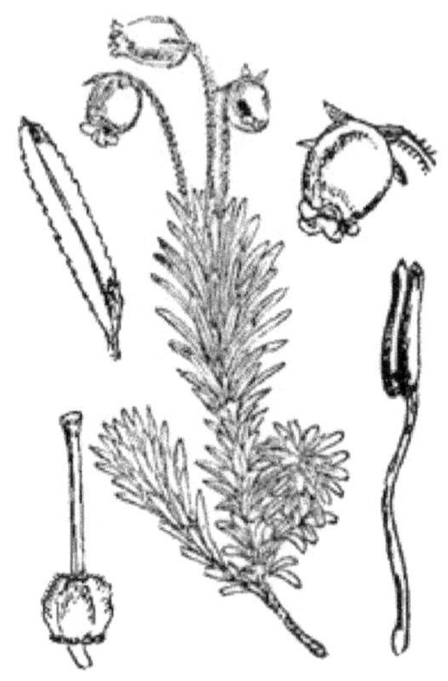

633. Menziesia cærulea.

634. Erica vulgaris.

635. Erica cinerea.

636. Erica Tetralix.

638. Erica carnea.

637. Erica ciliaris.

639. Erica vagans.

640. Pyrola uniflora.

641. Pyrola rotundifolia.

642. Pyrola media.

643. Pyrola minor.

644. Pyrola secunda.

645. Monotropa Hypopitys.

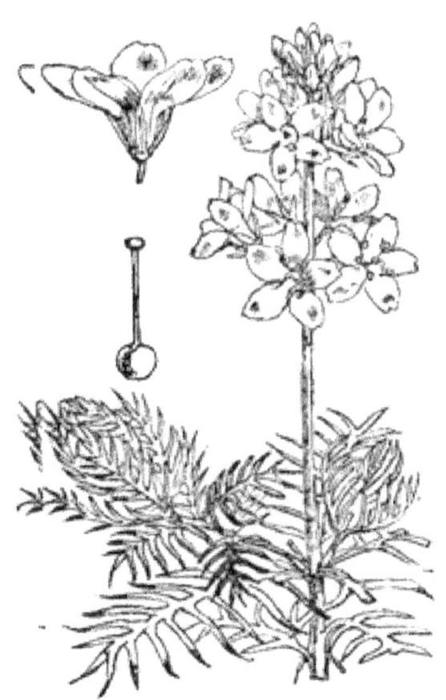

646. Hottonia palustris.

647. Primula vulgaris.

648. Primula veris.

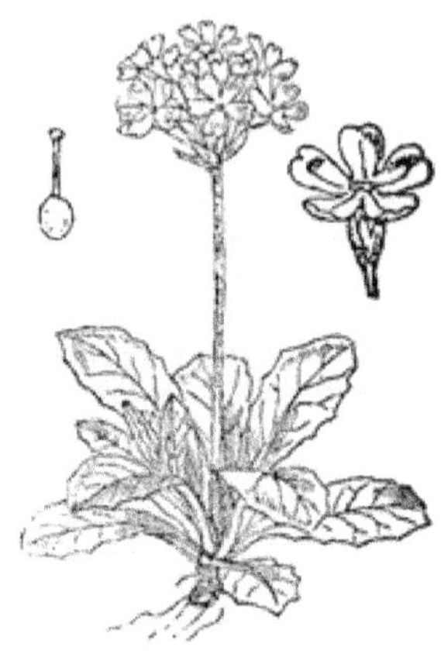

649. Primula farinosa.

650. Cyclamen europæum.

651. Lysimachia vulgaris.

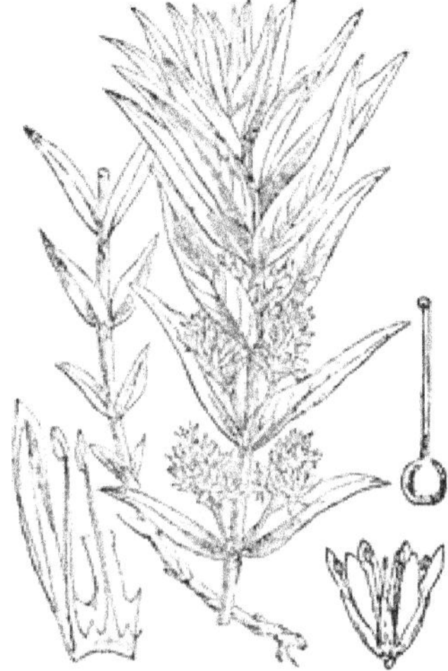

652. Lysimachia thyrsiflora.

654. Lysimachia nemorum.

653. Lysimachia nummularia.

655. Trientalis europæa.

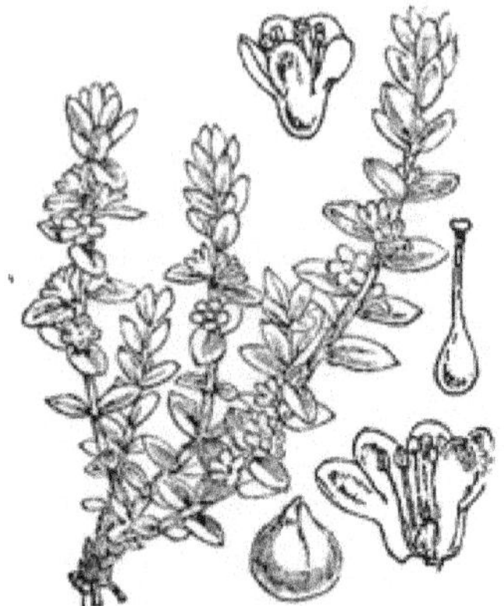

656. Glaux maritima.

657. Anagallis arvensis.

658. Anagallis tenella.

659. Centunculus minimus.

660. Samolus Valerandi

661. Pinguicula vulgaris.

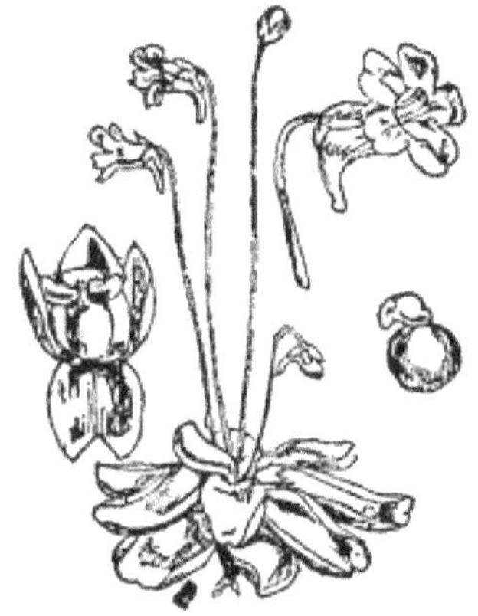

663. Pinguicula lusitanica.

662. Pinguicula alpina.

664. Utricularia vulgaris.

665. Utricularia minor.

666. Utricularia intermedia.

668. Ligustrum vulgare.

667. Fraxinus excelsior.

669. Vinca major.

670. Vinca minor.

671. Cicendia filiformis.

673. Erythræa centaurium.

672. Cicendia pusilla.

674. Gentiana pneumonanthe.

675. Gentiana verna.

676. Gentiana nivalis.

677. Gentiana amarella.

678. Gentiana campestris.

680. Menyanthes trifoliata.

679. Chlora perfoliata.

681. Limnanthemum nymphæoides.

682. Polemonium cæruleum.

683. Convolvulus arvensis.

684. Convolvulus sepium.

685, Convolvulus soldanella.

686. Cuscuta europæa.

687. Cuscuta epilinum.

688. Cuscuta epithymum.

689. Echium vulgare.

690. Echium violaceum.

691. Pulmonaria officinalis.

692. Mertensia maritima.

693. Lithospermum arvense.

694. Lithospermum officinale.

695. Lithospermum purpureo-cæruleum.

696. Myosotis palustris.

697. Myosotis sylvatica.

698. Myosotis arvensis.

699. Myosotis collina.

700. Myosotis versicolor.

701. Anchusa officinalis.

702. Anchusa sempervirens.

703. Lycopsis arvensis.

705. Symphytum tuberosum.

704. Symphytum officinale.

706. Borago officinalis.

707. Asperergo procumbens.

709. Cynoglossum montanum.

708. Cynoglossum officinale.

710. Dàtura stramonium.

711. Hyoscyamus niger.

712. Solanum dulcamara.

713. Solanum nigrum.

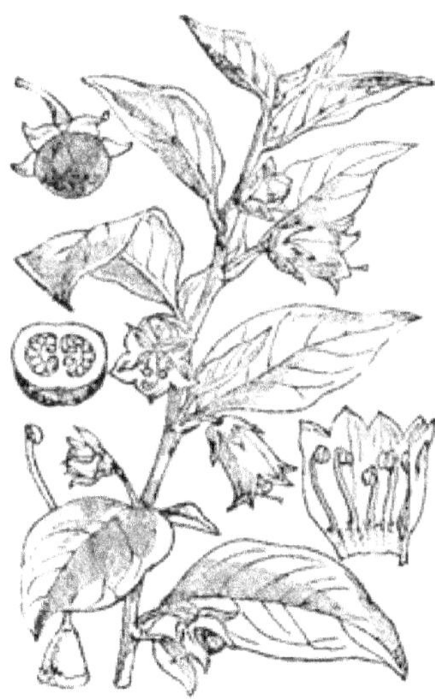

714. Atropa belladonna.

715. Orobanche major.

717. Orobanche rubra.

716. Orobanche caryophyllacea.

718. Orobanche elatior.

719. Orobanche minor.

721. Orobanche ramosa.

720. Orobanche cærulea.

722. Lathræa squamaria.

723. Verbascum thapsus.

724. Verbascum blattaria.

725. Verbascum virgatum.

726. Verbascum nigrum.

727. Verbascum Lychnitis.

729. Antirrhinum majus.

728. Verbascum pulverulentum.

730. Antirrhinum Orontium.

731. Linaria vulgaris.

732. Linaria repens.

733. Linaria Pelisseriana.

734. Linaria supina.

735. Linaria minor.

737. Linaria spuria.

36. Linaria Cymbalaria.

738. Linaria Elatine.

739. Scrophularia nodosa.

740. Scrophularia aquatica.

741. Scrophularia Scorodonia.

742. Scrophularia vernalis.

743. Minulus luteus.

745. Sibthorpia europæa.

744. Limosella aquatica.

746. Digitalis purpurea.

747. Veronica spicata.

748. Veronica saxatilis.

749. Veronica alpina.

750. Veronica serpyllifolia.

751. Veronica officinalis.

752. Veronica Anagallis.

753. Veronica Beccabunga.

754. Veronica scutellata.

755. Veronica montana.

757. Veronica hederæfolia.

756. Veronica Chamædrys.

758. Veronica agrestis.

759. Veronica Buxbaumii.

761. Veronica verna.

760. Veronica arvensis.

762. Veronica triphyllos.

763. Bartsia alpina.

764. Bartsia viscosa.

765. Bartsia Odontites.

766. Euphrasia officinalis.

767. Rhinanthus Crista-galli.

768. Pedicularis palustris.

769. Pedicularis sylvatica.

770. Melampyrum cristatum.

771. Melampyrum arvense.

772. Melampyrum pratense.

773. Melampyrum sylvaticum.

774. Salvia pratensis.

775. Salvia verbenaca.

776. Lycopus europæus.

777. Mentha sylvestris.

778. Mentha rotundifolia.

779. Mentha viridis.

780. Mentha piperita.

781. Mentha aquatica.

782. Mentha sativa.

783. Mentha arvensis.

784. Mentha Pulegium.

785. Thymus Serpyllum.

786. Origanum vulgare.

787. Calamintha Acinos.

788. Calamintha officinalis.

789. Calamintha Clinopodium.

790. Nepeta Glechoma.

791. Nepeta Cataria.

792. Prunella vulgaris.

793. Scutellaria galericulata.

794. Scutellaria minor.

795. Melittis Melissophyllum.

796. Marrubium vulgare.

797. Stachys Betonica.

798. Stachys germanica.

799. Stachys sylvatica.

800. Stachys palustris.

801. Stachys arvensis.

802. Galeopsis Ladanum.

803. Galeopsis ochroleuca.

804. Galeopsis Tetrahit.

805. Ballota nigra.

806. Leonurus Cardiaca.

807. Lamium amplexicaule.

808. Lamium purpureum.

809. Lamium album.

810. Lamium maculatum.

811. Lamium Galiobdolon.

813. Teucrium Scordium.

812. Teucrium Scorodonia.

814. Teucrium Botrys.

815. Teucrium Chamædrys.

816. Ajuga reptans.

817. Ajuga genevensis.

818. Ajuga Chamæpitys.

819. Verbena officinalis.

820. Statice Limonium.

821. Statice auriculifolia.

822. Statice reticulata.

823. Armeria vulgaris.

824. Armeria plantaginea.

825. Plantago major.

826. Plantago media.

827. Plantago lanceolata.

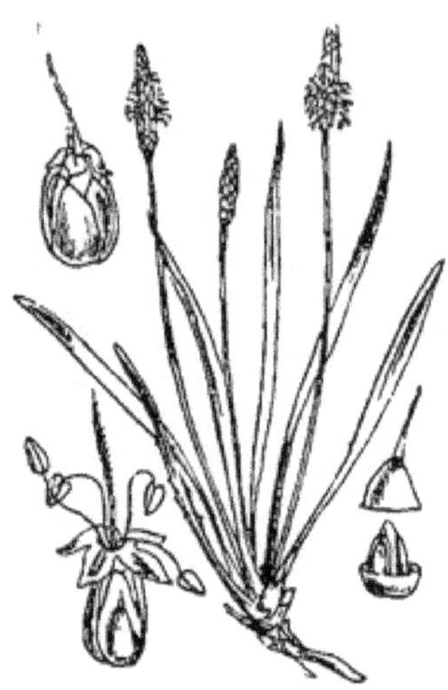

828. Plantago maritima.

829. Plantago Coronopus.

830. Littorella lacustris.

831. Corrigiola littoralis.

832. Herniaria glabra.

834. Scleranthus annuus.

833. Illecebrum verticillatum.

835. Scleranthus perennis.

836. Salicornia herbacea.

837. Suæda fruticosa.

839. Salsola Kali.

838. Suæda maritima.

840. Chenopodium Vulvaria.

841. Chenopodium polyspermum.

842. Chenopodium album.

843. Chenopodium glaucum.

844. Chenopodium rubrum.

845. Chenopodium urbicum.

846. Chenopodium murale.

847. Chenopodium hybridum.

848. Chenopodium Bonus-Henricus.

849. Beta maritima.

850. Atriplex portulacoides.

851. Atriplex pedunculata.

852. Atriplex hortensis.

853. Atriplex patula.

854. Atriplex rosea.

855. Rumex aquaticus.

856. Rumex crispus.

857. Rumex obtusifolius.

859. Rumex conglomeratus.

858. Rumex Hydrolapathum.

860. Rumex sanguineus.

861. Rumex pulcher.

862. Rumex maritimus.

863. Rumex Acetosa.

864. Rumex Acetosella.

865. Oxyria reniformis.

866. Polygonum aviculare.

867. Polygonum maritimum.

868. Polygonum Convolvulus.

869. Polygonum dumetorum.

870. Polygonum viviparum.

871. Polygonum Bistorta.

872. Polygonum amphibium.

873. Polygonum Persicaria.

874. Polygonum lapathifolium.

875. Polygonum Hydropiper.

876. Polygonum minus.

877. Daphne Mezereum.

878. Daphne Laureola.

879. Hippophae rhamnoides.

880. Thesium linophyllum.

881. Asarum europæum.

883 Euphorbia Helioscopia.

882. Euphorbia Peplis.

884. Euphorbia platyphyllos.

885. Euphorbia hiberna.

886. Euphorbia pilosa.

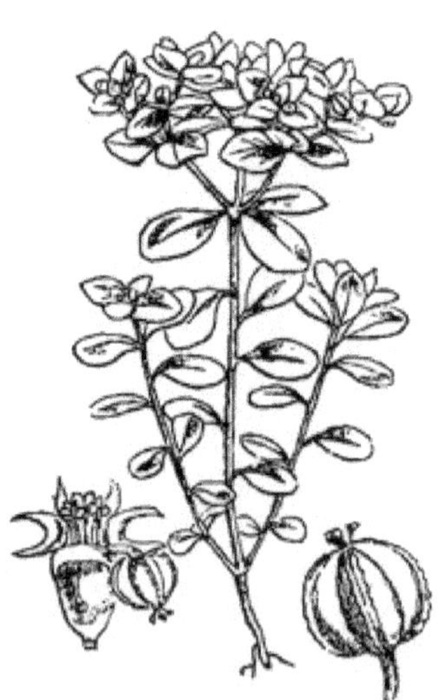

887. Euphorbia Peplus.

888. Euphorbia exigua.

889. Euphorbia Lathyris.

891. Euphorbia Paralias.

890. Euphorbia segetalis.

892. Euphorbia Esula.

893. Euphorbia amygdaloides.

894. Mercurialis perennis.

895. Mercurialis annua.

896. Buxus sempervirens.

897. Empetrum nigrum.

899. Callitriche aquatica.

898. Ceratophyllum demersum.

900. Urtica urens.

901. Urtica pilulifera.

902. Urtica dioica.

903. Parietaria officinalis.

904. Humulus Lupulus.

905. Ulmus montana.

907. Myrica Gale.

906. Ulmus campestris.

908. Alnus glutinosus.

909. Betula alba.

910. Betula nana.

911. Carpinus Betulus.

912. Corylus Avellana.

913. Fagus sylvatica.

914. Quercus Robur.

915. Salix pentandra.

916. Salix fragilis.

917. Salix alba.

918. Salix amygdalina.

919. Salix purpurea.

920. Salix viminalis.

921. Salix Caprea.

922. Salix aurita.

923. Salix phylicifolia.

924. Salix repens.

925. Salix Lapponum.

927. Salix myrsinites.

928. Salix reticulata.

926. Salix lanata.

929. Salix herbacea.

930. Populus alba.

931. Populus tremula.

932. Populus nigra.

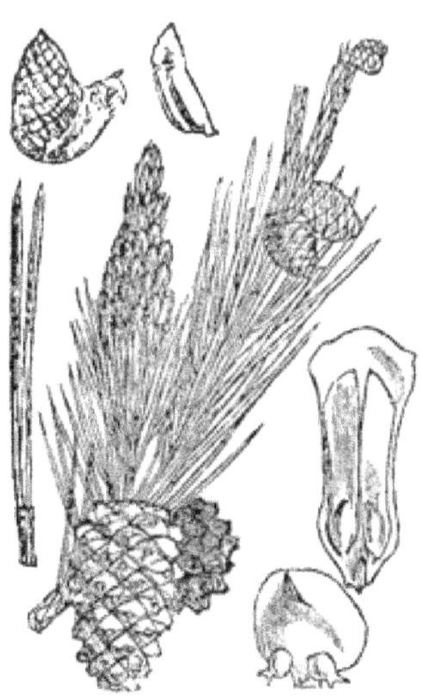

933. Pinus sylvestris.

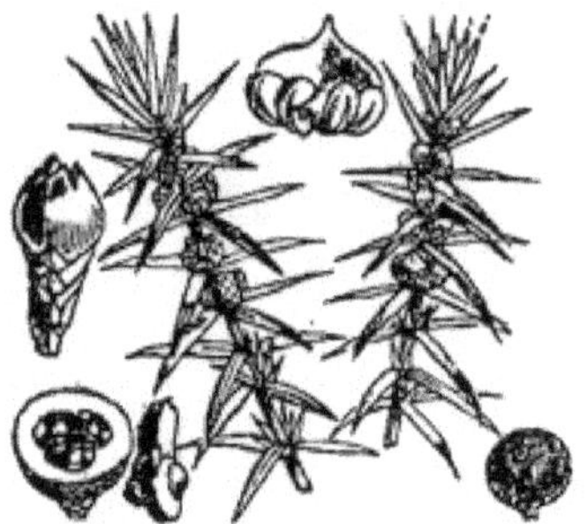

934. Juniperus communis.

935. Taxus baccata.

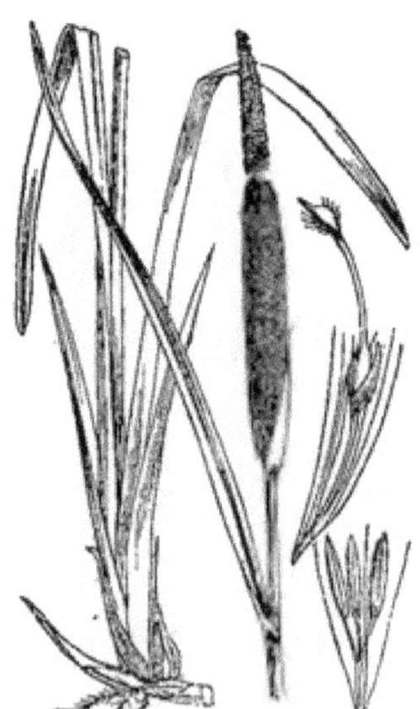

936. Typha latifolia.

937. Typha angustifolia.

938. Sparganium ramosum.

939. Sparganium simplex.

940. Sparganium minimum.

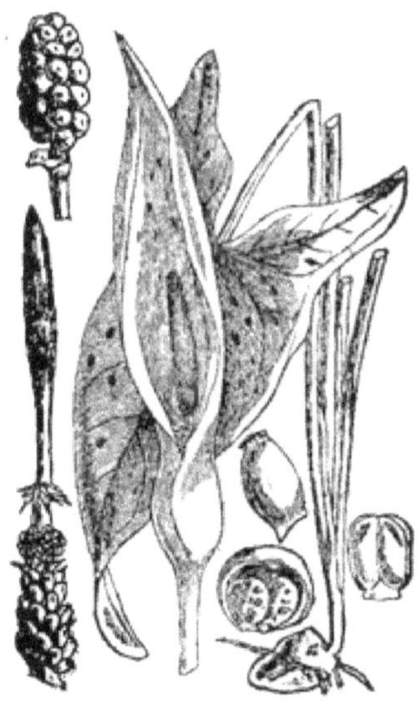
941. Arum maculatum.

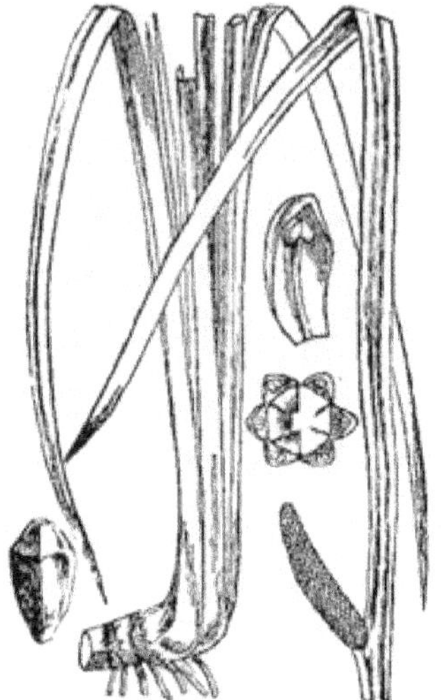

942. Acorus calamus.

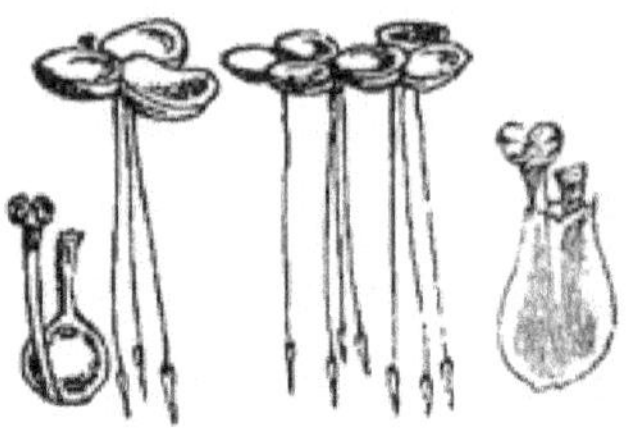

945. Lemna gibba.

943. Lemna trisulca.

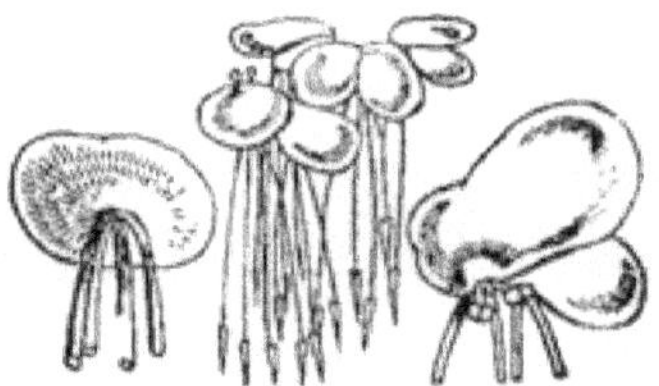

946. Lemna polyrrhiza.

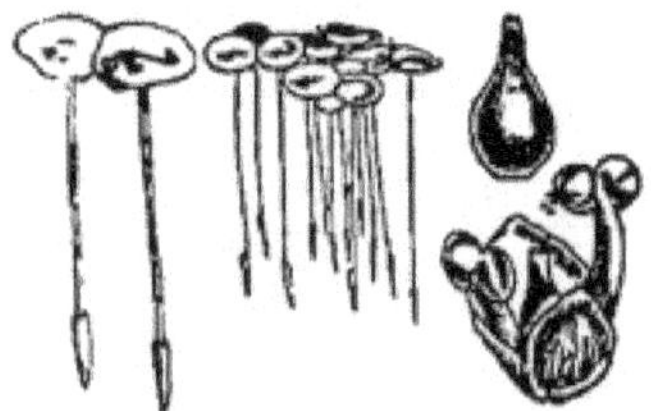

944. Lemna minor.

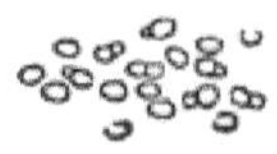

947. Lemna arrhiza.

948. Zostera marina.

949. Zostera nana.

950. Naias flexilis.

951. Zannichellia palustris.

952. Ruppia maritima.

953. Potamogeton natans.

954. Potamogeton heterophyllus.

955. Potamogeton lucens.

956. Potamogeton prælongus.

957. Potamogeton perfoliatus.

958. Potamogeton crispus.

959. Potamogeton densus.

960. Potamogeton obtusifolius.

961. Potamogeton acutifolius.

962. Potamogeton pusillus.

963. Potamogeton pectinatus.

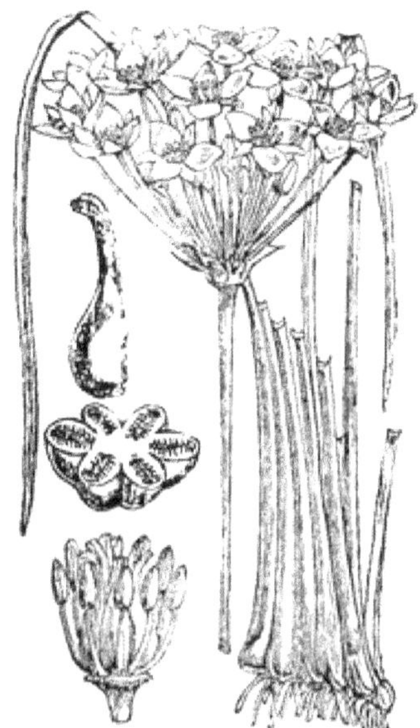

964. Butomus umbellatus.

966. Alisma Plantago.

965. Sagittaria sagittifolia.

967. Alisma ranunculoides.

968. Alisma natans.

969. Damasonium stellatum.

970. Scheuchzeria palustris.

971. Triglochin palustre.

972. Triglodrin maritimum.

973. Elodea canadensis.

974. Hydrocharis Morsus-ranæ.

975. Stratiotes aloides.

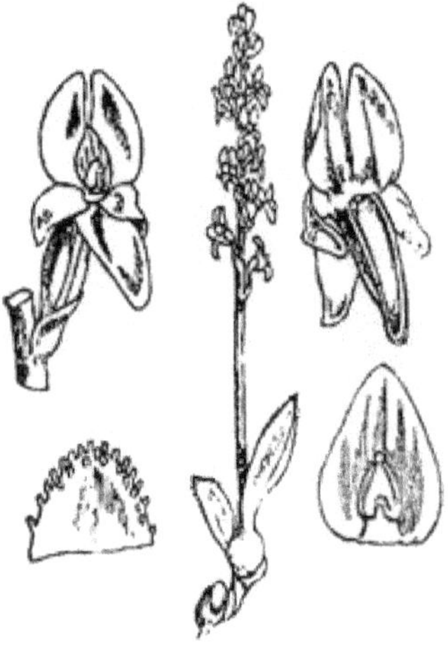

976. Malaxis paludosa.

978. Corallorhiza innata.

977. Liparis Loeselii.

979. Epipactis latifolia.

980. Epipactis palustris.

981. Cephalanthera grandiflora.

982. Cephalanthera ensifolia.

983. Cephalanthera rubra.

984. Listera ovata.

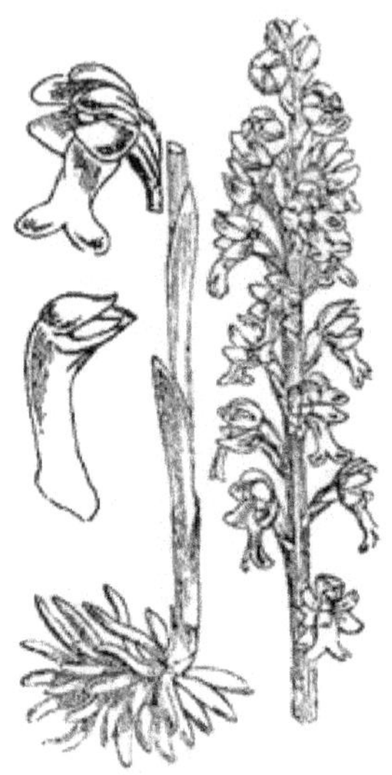

986. Neottia Nidus-avis.

985. Listera cordata.

987. Epipogium aphyllum.

988. Spiranthes autumnalis.

989. Spiranthes æstivalis.

990. Spiranthes cernua.

991. Goodyera repens.

992. Orchis Morio.

993. Orchis militaris.

994. Orchis ustulata.

995. Orchis intacta.

996. Orchis mascula.

997. Orchis laxiflora.

998. Orchis maculata.

999 Orchis latifolia.

1000. Orchis hircina.

1001. Orchis pyramidalis.

1002. Orchis conopsea.

1003. Habenaria bifolia.

1004. Habenaria albida

1005. Habenaria viridis.

1006. Aceras anthropophora.

1007. Herminium Monorchis.

1008. Ophrys apifera.

1009. Ophrys aranifera.

1010. Ophrys muscifera.

1011. Cypripedium Calceolus.

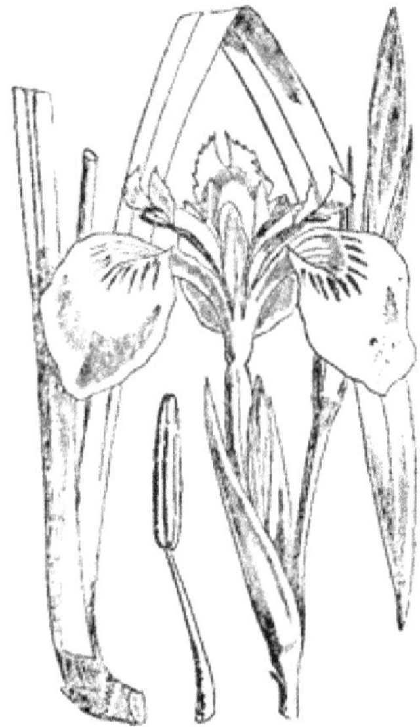
1012. Iris Pseudacorus.

1013. Iris fœtidissima.

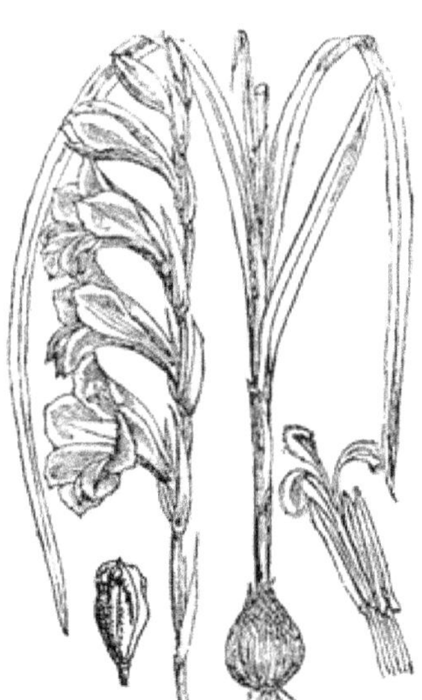
1014. Gladiolus communis.

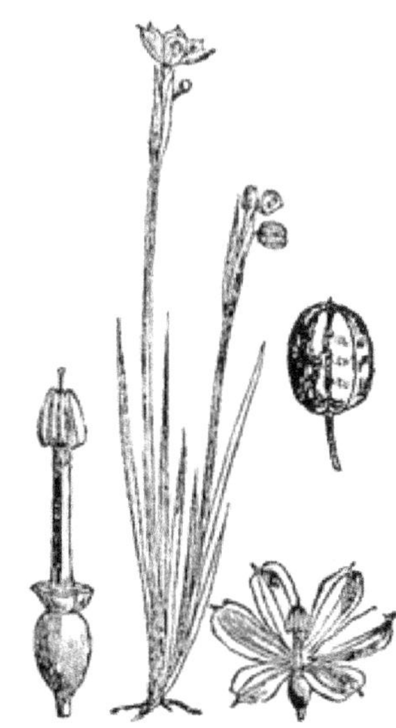
1015. Sisyrinchum Bermudiana.

1016. Trichonema Bulbocodium.

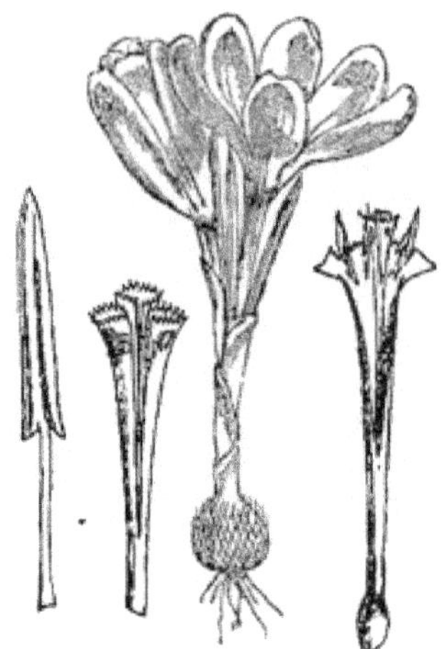

1017. Crocus vernus.

1018. Crocus nudiflorus.

1019. Narcissus Pseudonarcissus.

1020. Narcissus biflorus.

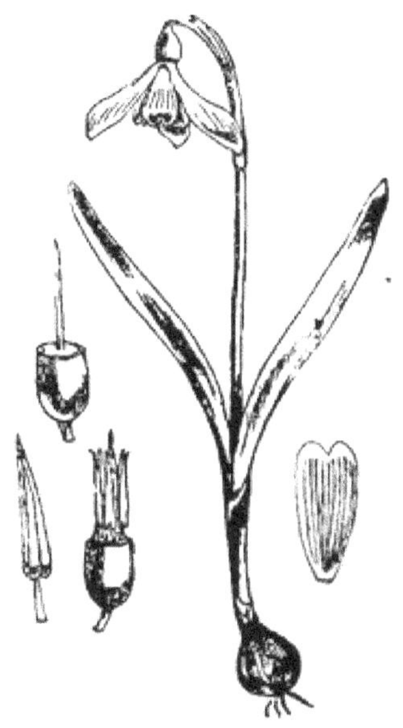

1021. Galanthus nivalis.

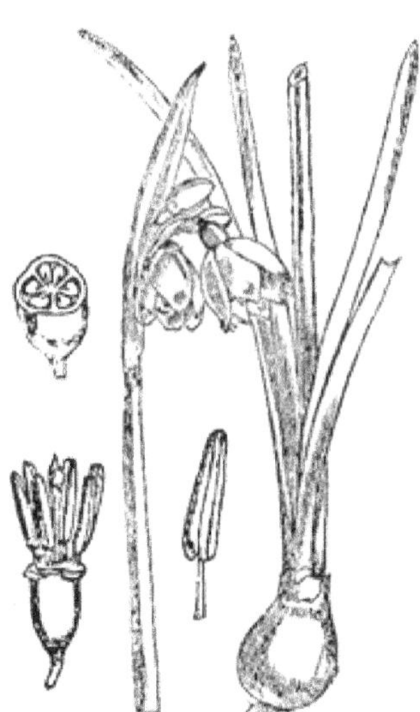

1022. Leucoium æstivum.

1023. Tamus communis.

1024. Paris quadrifolia.

1025. Polygonatum verticillatum.

1026. Polygonatum multiflorum.

1027. Polygonatum officinale.

1028. Convallaria majalis.

1029. Smilacina bifolia.

1030. Asparagus officinalis.

1031. Ruscus aculeatus.

1032. Fritillaria Meleagris.

1033. Tulipa sylvestris.

1034. Lloydia serotina.

1035. Gagea lutea.

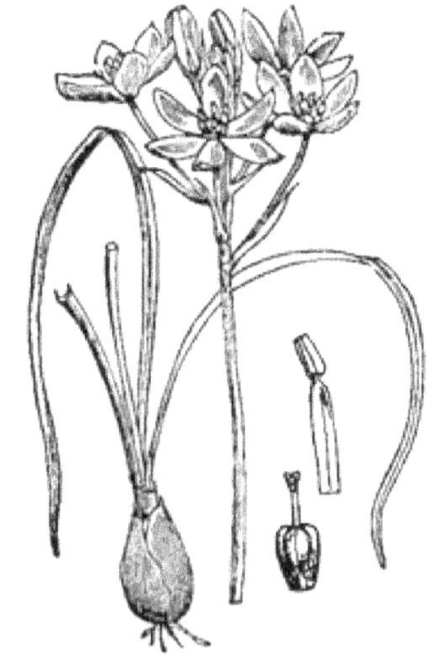

1036. Ornithogalum umbellatum.

1037. Ornithogalum nutans.

1038. Ornithogalum pyrenaicum.

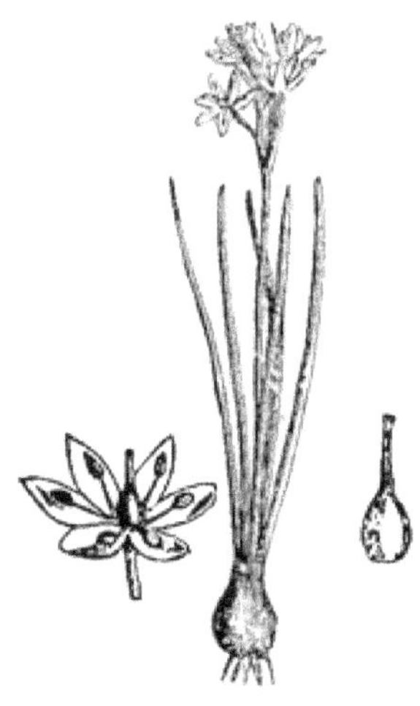

1039. Scilla verna.

1040. Scilla autumnalis.

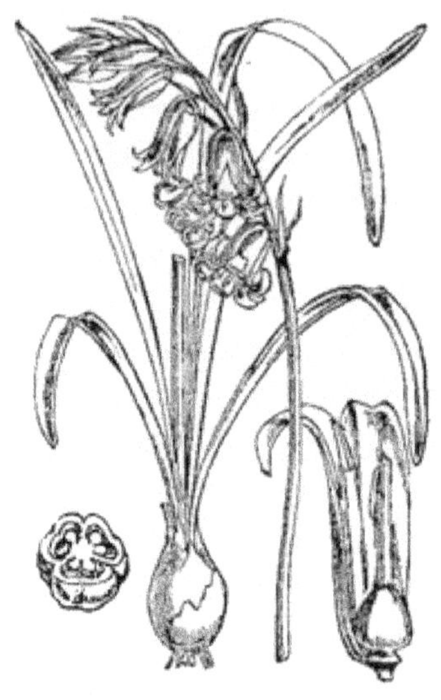

1041. Scilla nutans.

1042. Muscari racemosum.

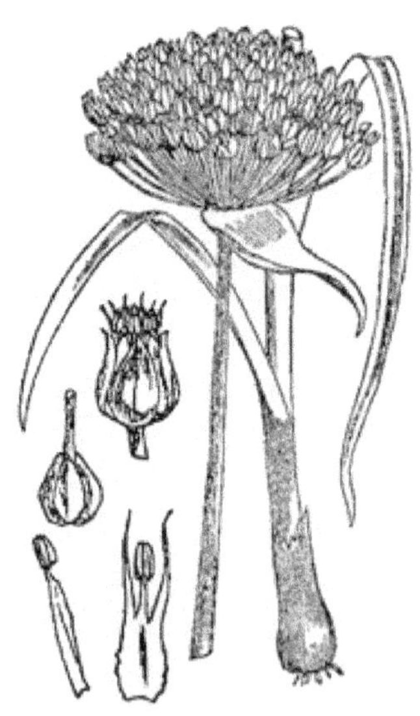

1043 Allium Ampeloprasum.

1044. Allium Scorodoprasum.

1045. Allium oleraceum.

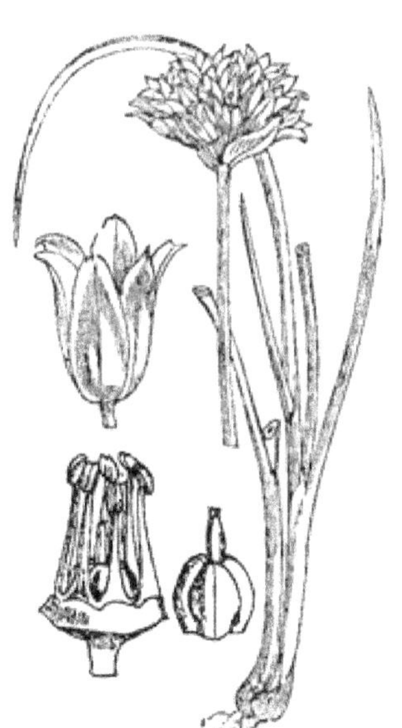

1046. Allium Schœnoprasum.

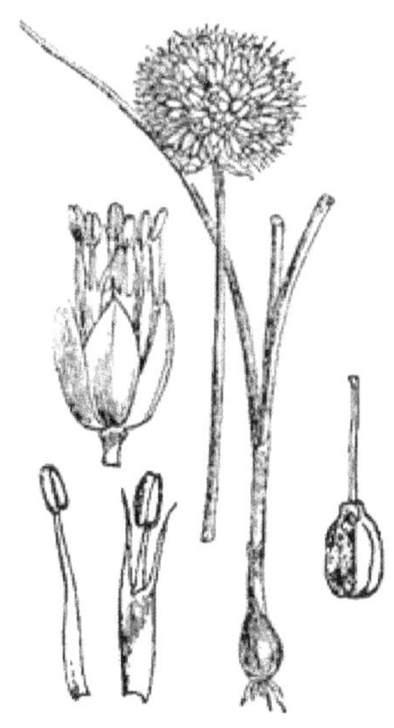

1047. Allium sphærocephalum.

1048. Allium vineale.

1049. Allium ursinum.

1050. Allium triquetrum.

1051. Simethis bicolor.

1052. Narthecium ossifragum.

1054. Colchicum autumnale.

1053. Tofieldia palustris.

1055. Juncus communis.

1056. Juncus glaucus.

1057. Juncus filiformis.

1058. Juncus balticus.

1059. Juncus articulatus.

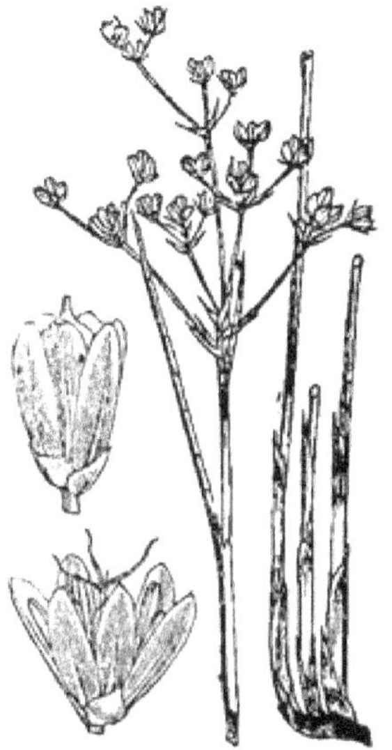
1060. Juncus obtusiflorus.

1061. Juncus compressus.

1062. Juncus squarrosus.

1063. Juncus bufonius.

1064. Juncus pygmæus.

1065. Juncus capitatus.

1066. Juncus maritimus.

1067. Juncus acutus.

1068. Juncus trifidus.

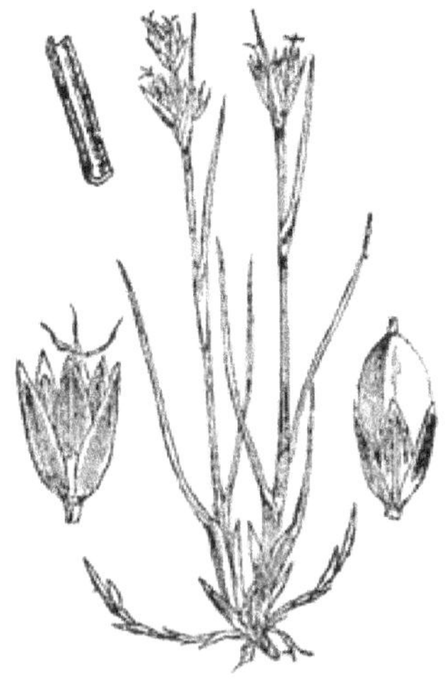

1069. Juncus castaneus.

1070. Juncus biglumis.

1071. Luzula pilosa.

1072. Luzula sylvatica.

1074. Luzula campestris.

1073. Luzula arcuata.

1075. Luzula spicata.

1076. Eriocaulon septangulare.

1078. Cyperus fuscus.

1077. Cyperu longus.

1079. Schœnus nigricans.

1080. Cladium Mariscus.

1081. Rhynchospora fusca.

1082. Rhynchospora alba.

1083. Blysmus compressus.

1084. Blysmus rufus.

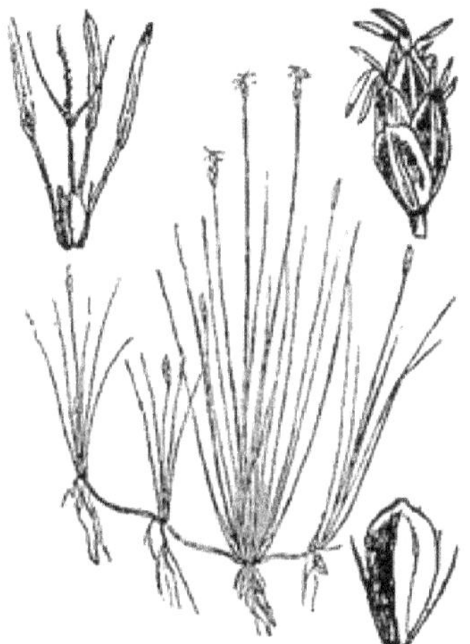

1085. Scirpus acicularis.

1086. Scirpus parvulus.

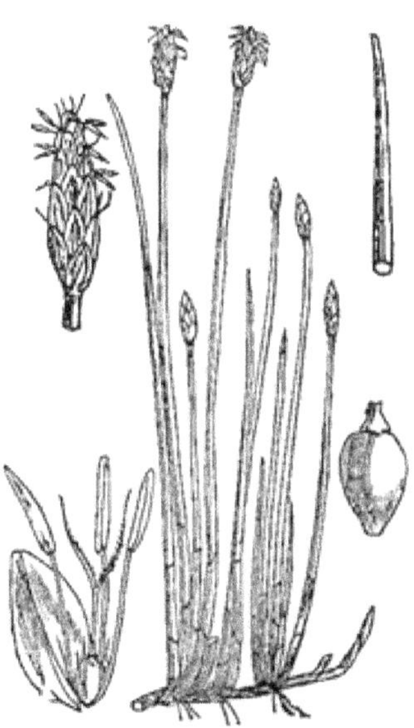

1087. Scirpus palustris.

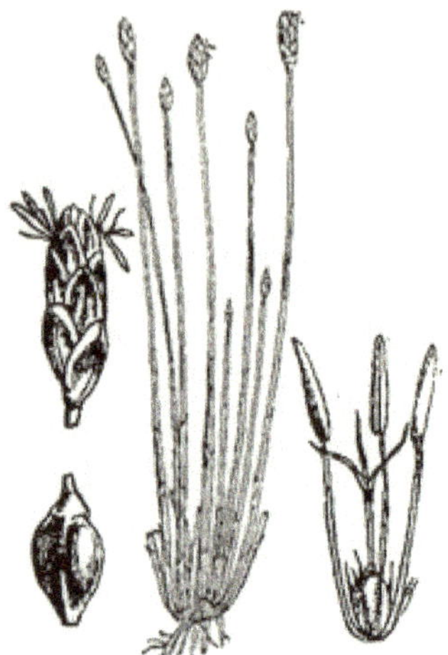

1088. Scirpus multicaulis.

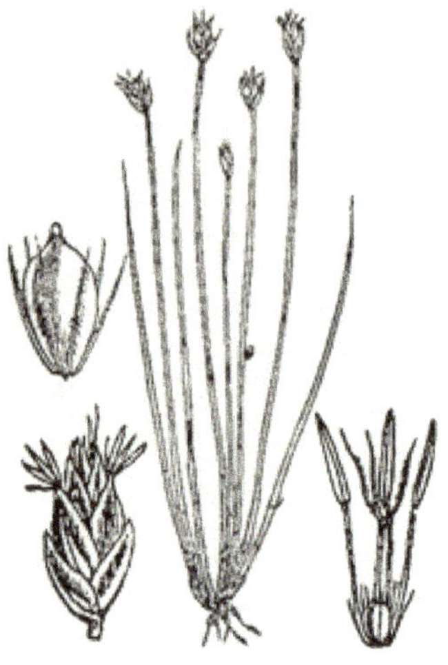

1089. Scirpus pauciflorus.

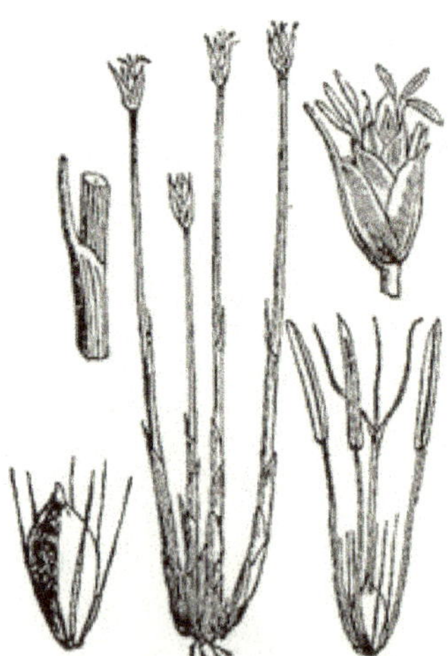

1090. Scirpus cæspitosus.

1091. Scirpus fluitans.

1092. Scirpus setaceus.

1093. Scirpus riparius.

1094. Scirpus Holoschœnus.

1095. Scirpus pungens.

1096. Scirpus triqueter.

1097. Scirpus lacustris.

1098. Scirpus maritimus.

1099. Scirpus sylvaticus.

1100. Eriophorum alpinum.

1101. Eriophorum vaginatum.

1102. Eriophorum polystachyum.

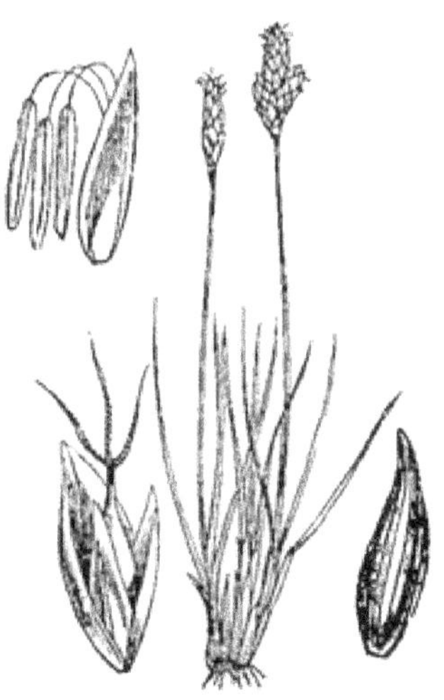

1103. Kobresia caricina.

1104. Carex dioica.

1105. Carex pulicaris.

1106. Carex rupestris.

1107. Carex panciflora.

1108. Carex leporina.

1109. Carex lagopina.

1110. Carex elongata.

1111. Carex stellulata.

1112. Carex canescens.

1113. Carex remota.

1114. Carex axillaris.

1115. Carex paniculata.

1116. Carex vulpina.

1118. Carex arenaria.

1117. Carex muricata.

1119. Carex divisa.

1120. Carex incurva.

1121. Carex saxatilis.

1122. Carex cæspitosa.

1123. Carex acuta.

1124. Carex alpina.

1125. Carex Buxbaumii.

1126. Carex atrata.

1127. Carex humilis.

1128. Carex digitata.

1129. Carex præcox.

1130. Carex montana.

1131. Carex pilulifera.

1132. Carex tomentosa.

1133. Carex filiformis.

1134. Carex hirta.

1135. Carex pallescens.

1136. Carex extensa.

1137. Carex flava.

1138 Carex distans.

1139. Carex punctata.

1140. Carex panicea.

1141. Carex capillaris.

1142. Carex limosa.

1143. Carex glauca.

1144. Carex sylvatica.

1145. Carex strigosa.

1146. Carex Pseudocyperus.

1147. Carex pendula.

1148. Carex ampullacea.

1149. Carex vesicaria.

1150. Carex paludosa.

1151. Leersia oryzoides.

1152. Milium effusum.

1153. Panicum sanguinale.

1154. Panicum glabrum.

1155. Panicum verticillatum.

1156. Panicum glaucum.

1157. Panicum viride.

1159. Hierochloe borealis.

1158. Panicum Crus-galli.

1160. Anthoxanthum odoratum.

1161. Phalaris canariensis.

1162. Digraphis arundinacea.

1163. Phleum pratense.

1164. Phleum alpinum.

1165. Phleum Bœhmeri.

1166. Phleum asperum.

1167. Phleum arenarium.

1168. Alopecurus agrestis.

1169. Alopecurus pratensis.

1170. Alopecurus geniculatus.

1171. Alopecurus alpinus.

1172. Chamagrostis minima.

1173. Lagurus ovatus.

1175. Polypogon littoralis.

1174. Polypogon monspeliensis.

1176. Agrostis alba.

1177. Agrostis canina.

1178. Agrostis cetacea.

1179. Agrostis Spica-venti.

1180. Gastridium lendigerum.

1181. Psamma arenaria.

1182. Calamagrostis Epigeios.

1183. Calamagrostis lanceolata.

1184. Calamagrostis stricta.

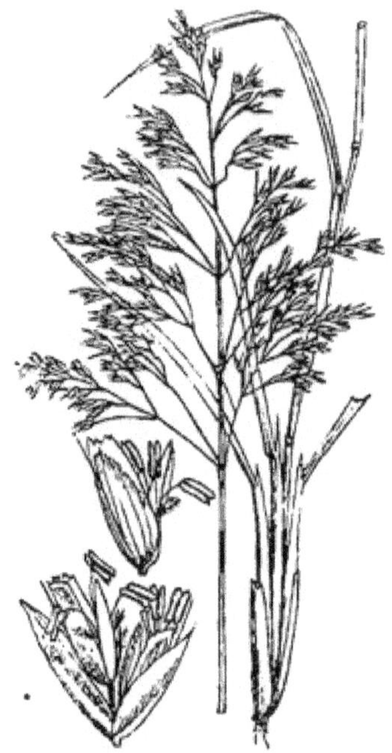
1185. Aira cæspitosa.

1187. Aira canesceus.

1186. Aira flexuosa.

1188. Aira præcox.

1189. Aira caryophyllea.

1190. Avena fatua.

1191. Avena pratensis.

1192. Avena flavescens.

1193. Arrhenatherum avenaceum.

1195. Holcus mollis.

1194. Holcus lanatus.

1196. Cynodon Dactylon.

1197. Spartina stricta.

1199 Nardus stricta.

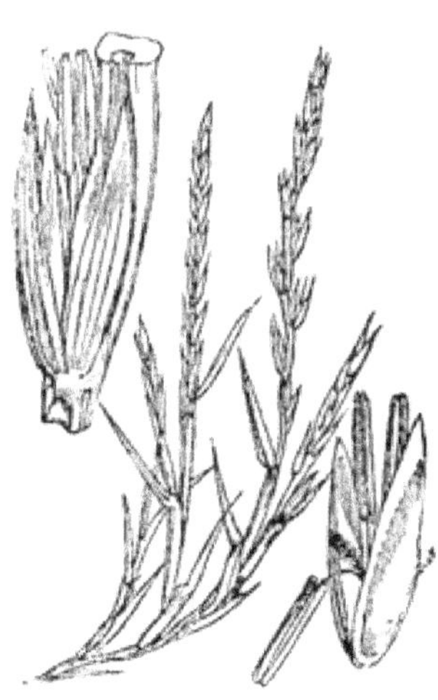

1198. Lepturus incurvatus.

1200. Elymus arenarius.

1201. Hordeum sylvaticum.

1202. Hordeum pratense.

1203. Hordeum murinum.

1204. Hordeum maritimum.

1206. Agropyrum caninum.

1205. Agropyrum repens.

1207. Lolium perenne.

1208. Lolium temulentum.

1209. Brachypodium sylvaticum.

1210. Brachypodium pinnatum.

1211. Bromus erectus.

1212. Bromus asper.

1213. Bromus sterilis.

1215. Bromus madritensis.

1214. Bromus maximus.

1216. Bromus arvensis.

1218. Festuca ovina.

1217. Bromus giganteus.

1219. Festuca elatior.

1220. Festuca sylvatica.

1222. Festuca uniglumis.

1221. Festuca Myurus.

1223. Dactylis glomerata.

1224. Cynosurus cristatus.

1226. Briza media.

1225. Cynosurus echinatus.

1227. Briza minor.

1228. Poa aquatica.

1229. Poa fluitans.

1230. Poa maritima.

1231. Poa distans.

1232. Poa procumbens.

1233. Poa rigida.

1234. Poa loliacea.

1235. Poa annua.

1236. Poa compressa.

1237. Poa pratensis.

1238. Poa trivialis.

1239. Poa memoralis.

1240. Poa laxa.

1241. Poa alpina.

1242. Poa bulbosa.

1243. Catabrosa aquatica.

1244. Molinia cærulea.

1245. Melica nutans.

1247. Triodia decumbens.

1246. Melica uniflora.

1248. Kœleria cristata.

1249. Sesleria cærulea.

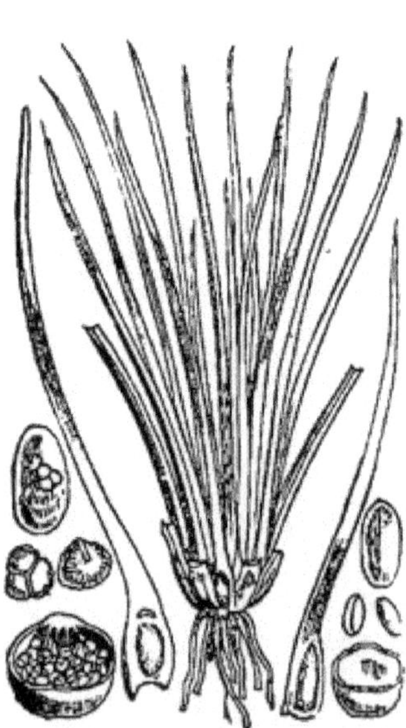

1251. Isoetes locustris.

1250. Arundo Phragmites.

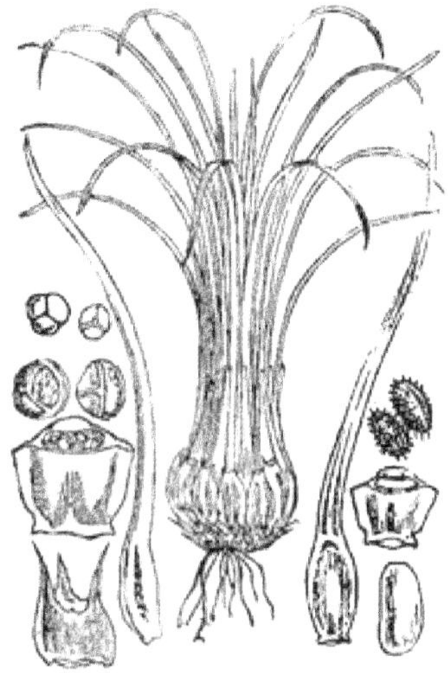

1252. Isoetes Duriæe.

1253. Lycopodium clavatum.

1255. Lycopodium alpinum.

1254. Lycopodium annotinum.

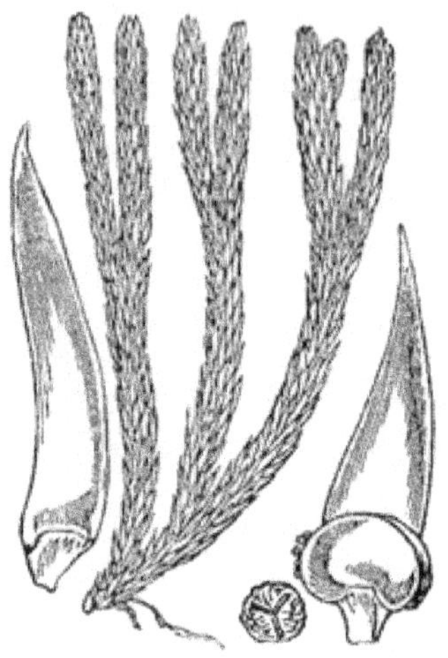

1256. Lycopodium Selago.

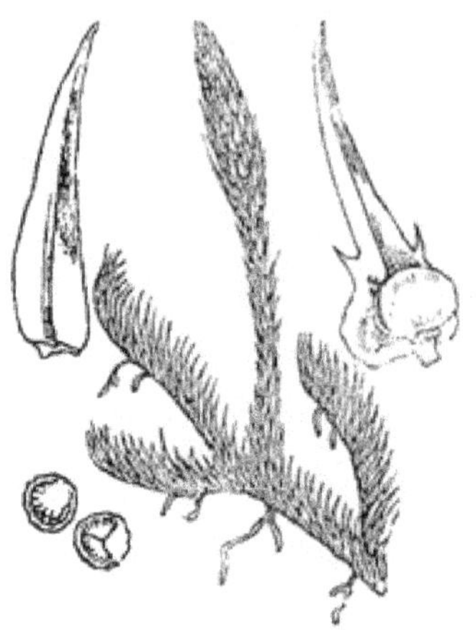

1257. Lycopodium inundatum.

1258. Lycopodium selaginoides.

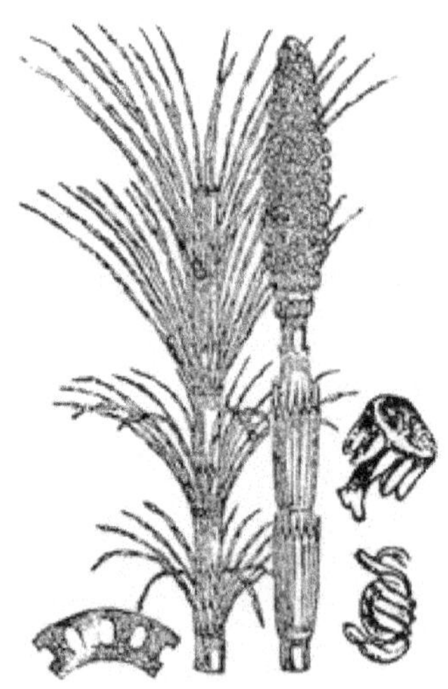

1259. Equisetum Telmateia.

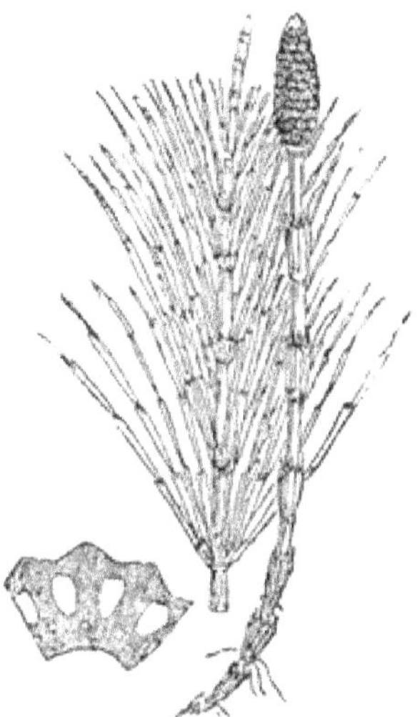

1260. Equisetum arvense.

1261. Equisetum sylvaticum.

1262. Equisetum pratense.

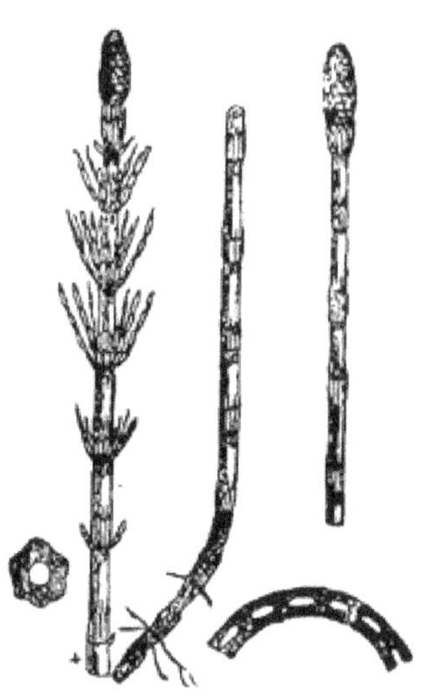

1263. Equisetum limosum.

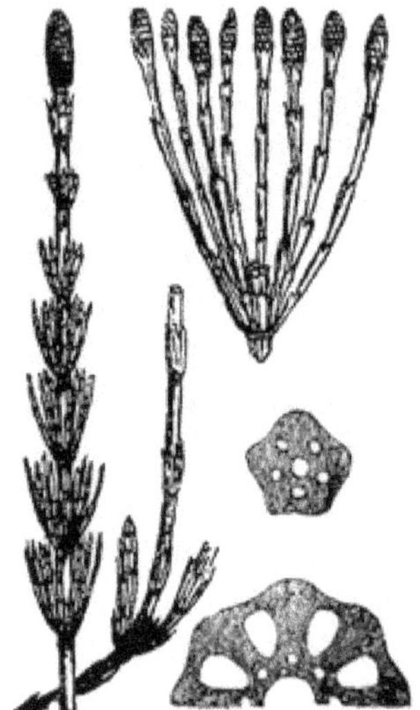

1264. Equisetum palustre.

1265. Equisetum hyemale.

1266. Equisetum ramosum.

1267. Equisetum variegatum.

1268. Pilularia globulifera.

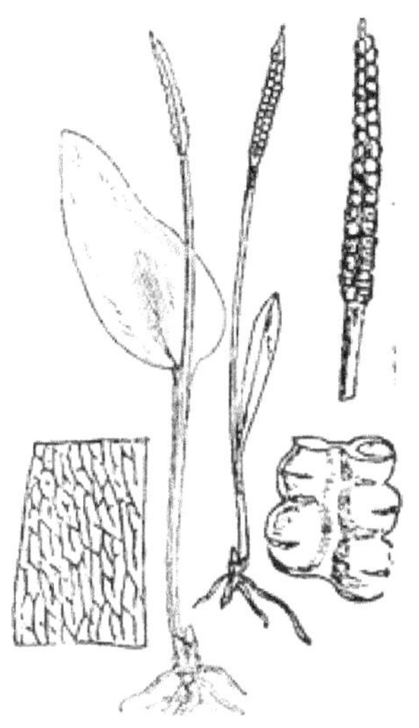
1269. Ophioglossum vulgatum.

1271. Osmunda regalis.

1270. Botrychium Lunaria.

1272. Polypodium vulgare.

1273. Polypodium Phegopteris.

1274. Polypodium alpestre.

1275. Polypodium Dryopteris.

1276. Allosorus crispus.

1277. Grammitis leptophylla.

1278. Aspidium Lonchitis.

1279. Aspidium aculeatum.

1280. Aspidium Thelypteris.

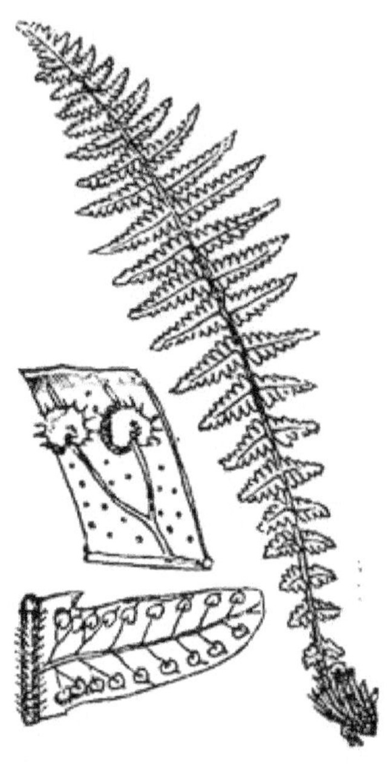

1281. Aspidium Oreopteris.

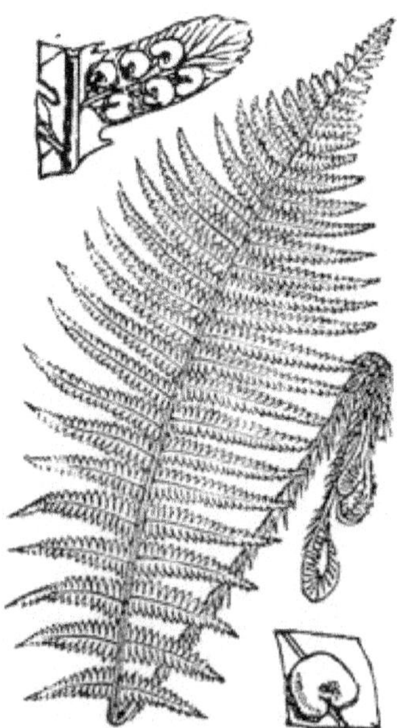

1282. Aspidium Filix-mas.

1283. Aspidium cristatum.

1284. Aspidium spinulosum.

1286. Asplenium Filix-fœmina.

1285. Aspidium rigidum.

1287. Asplenium fontanum.

1288. Asplenium lanceolatum.

1290. Asplenium Trichomanes.

1289. Asplenium marinum.

1291. Asplenium viride.

1292. Asplenium Adiantum-nigrum.

1294. Asplenium germanicum.

1293. Asplenium Ruta-muraria.

1295. Asplenium septentrionale.

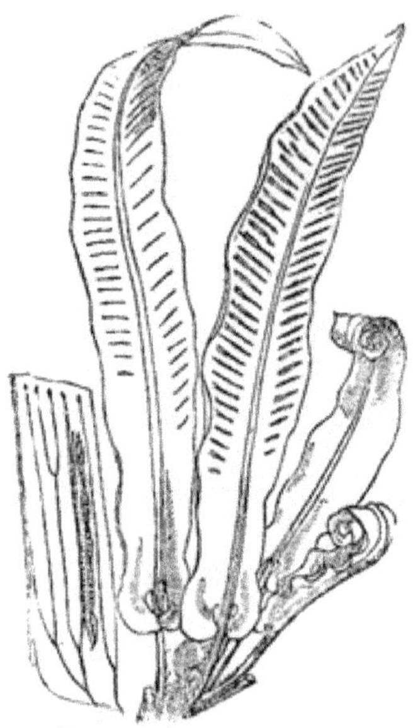

1296. Scolopendrium vulgare.

1297. Ceterach officinarum.

1298. Blechnum Spicant.

1299. Pteris aquilina.

1300. Adiantum Capillus-Veneris.

1301. Cystopteris fragilis.

1302. Cystopteris montana.

1303. Woodsia ilvensis.

1304. Trichomanes radicans.

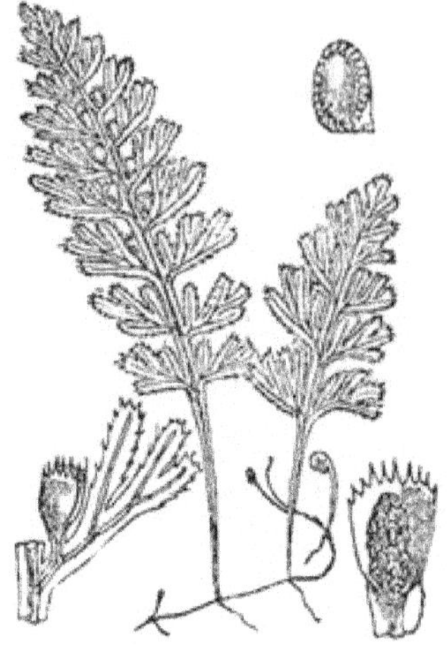

1305. Hymenophyllum tunbridgense.

1306. Hymenophyllum unilaterale.

INDEX.

Synonyms are in italics.

FIG.
Acer campestre . . 219
pseudo-platanus 220
Aceras anthropophora 1006
Achillea millefolium . 530
ptarmica . . 529
Aconitum napellus . 30
Acorus calamus . . 942
Actæa spicata 31
Actinocarpus
damasonium 969
stellatum . . 969
Adiantum
capillus-veneris . 1300
Adonis autumnalis . 7
Adoxa moschatellina 458
Ægopodium podagraria 406
Æthusa cynapium 424
Agrimonia eupatoria . 326
odorata . . . 326
Agropyrum caninum 1206
repens . . . 1205
Agrostemma
githago . . . 141
Agrostis alba . 1176
australis. . 1180
canina . . 1177
setacea . . 1178
spica-venti . . 1179
vulgaris . . 1176
Aira cæspitosa . . 1185
canescens . . 1187
caryophyllea . 1189
flexuosa . . . 1186
præcox . . . 1188
Ajuga chamæpitys . 818
genevensis . . 817
pyramidalis . 817
reptans . . . 816
Alchemilla alpina . 322
arvensis . . 323
conjuncta . . 322
vulgaris . . 321
Alisma natans 968
plantago . . . 966
ranunculoides . . 967
Alliaria officinalis . 72
Allium ampeloprasum 1043
oleraceum . . 1045
schœnoprasum . 1046
scorodoprasum . 1044
sphærocephalum . 1047
triquetrum . . 1050
ursinum . . 1049
vineale . . . 1048
Allosorus crispus. . 1276
Alnus glutinosus . 908
Alopecurus agrestis 1168
alpinus . . . 1171
geniculatus . . 1170
pratensis . . 1169
stricta . . . 150
Alsine
tenuifolia . 151

FIG.
Alsine
verna . . 149
Althæa hirsuta . . 198
officinalis . . . 197
Alyssum calycinum . 86
maritimum. 87
Anacharis
alsinastrum. . 973
canadensis . . 973
Anagallis arvensis 657
tenella . . 658
Anchusa *arvensis* . 703
officinalis. . 701
sempervirens . . 702
Andromeda polifolia . 630
Anemone nemorosa . 6
pulsatilla . . 5
Angelica sylvestris . 431
Antennaria dioica . 509
margaritacea . . 510
Anthemis arvensis 526
cotula . 525
nobilis . . . 527
tinctoria . . 528
Anthoxanthum
odoratum. . . 1160
Anthriscus sylvestre . 442
vulgaris . . . 443
Anthyllis vulneraria . 266
Antirrhinum majus 729
orontium . . . 730
Apargia autumnalis . 580
hispidus . . 579
Apera spica-venti . 1179
Apium graveolens . 401
inundatum . 403
nodiflorum . . 402
Aquilegia vulgaris . 28
Arabis ciliata . . 59
hirsuta . . . 58
perfoliata. 56
petræa . . 62
stricta . . . 61
thaliana . . 60
turrita . . 57
Arbutus unedo . 627
Arctium lappa . . 552
Arctostaphylos alpina. 629
uva-ursi. . . 628
Arenaria *cherleria* . 148
ciliata . . . 154
peploides . . 152
serpyllifolia . . 153
tenuifolia. 151
trinervis . . 155
uliginosa . . 150
verna . 149
Armeria plantaginea . 824
vulgaris . . 823
Armoracia
amphibium . 55
rusticana . 84
Arnoseris pusilla. . 607

FIG.
Arrhenatherum
avenaceum . 1193
Artemisia absinthium. 536
campestris . 533
maritima . . 534
vulgaris . . . 535
Arum maculatum 941
Arundo phragmites . 1250
Asarum europæum 881
Asparagus officinalis . 1030
Asperugo procumbens 707
Asperula cynanchica . 479
odorata . . . 478
Aspidium aculeatum . 1279
cristatum. . . 1283
filix-mas . . . 1282
lobatum . 1279
lonchitis . . 1278
oreopteris . . 1281
rigidum . . . 1285
spinulosum . . 1284
thelypteris . 1280
Asplenium
adiantum-nigrum 1292
ceterach . . . 1297
filix-fœmina . . 1286
fontanum. . 1287
germanicum . . 1294
lanceolatum . 1288
marinum . . . 1289
ruta-muraria . 1293
septentrionale . . 1295
trichomanes . . 1290
viride . . . 1291
Aster linosyris . . 496
tripolium . . 495
Astragalus alpinus 268
glycyphyllos . 269
hypoglottis . . 267
Astrantia major . . 397
Athyrium filix-
fœmina . . . 1286
Atriplex hortensis . 852
laciniata. . 854
nitens . . . 852
patula . 853
pedunculata . . 851
portulacoides . . 850
rosea . . . 854
Atropa belladonna . 714
Avena *elatior* . 1193
fatua . . . 1190
flavescens . . 1192
pratensis . . . 1191
Azalia procumbens 631

Ballota *fœtida* . 805
nigra . 805
ruderalis. . 805
Barbarea vulgaris . 51
Bartsia alpina . 763
odontites . . . 765
viscosa . . 764

FIG.
Bellis perennis . . 501
Berberis vulgaris 33
Beta maritima . 849
Betonica officinalis 797
Betula alba. . 909
glutinosa. . 909
nana . . 910
Bidens cernua 518
tripartita . . 519
Blechnum *boreale* . 1298
spicant . . . 1298
Blysmus compressus 1083
rufus. . 1084
Borago officinalis 706
Borkhausia fœtida 594
taraxicifolia . 593
Botrychium lunaria . 1270
Brachypodium
pinnatum . . 1210
sylvaticum . . 1209
Brassica adpressa . 83
alba 80
brevipes . . . 76
campestris . 79
monensis . . 77
muralis 76
nigra . 82
oleracea . 78
polymorpha 79
sinapistrum . . 81
tenuifolia . 75
Briza media . 1226
minor . . 1227
Bromus arvensis. . 1216
asper . . 1212
commutatus 1216
diandrus . . 1214
erectus . . 1211
giganteus . . 1217
madritensis . 1215
maximus . . 1214
sterilis . 1213
Bryonia dioica . . 356
Bunium bulbocastanum 411
flexuosum 440
Bupleurum aristatum. 417
falcatum . . 419
rotundifolium . 416
tenuissimum . 418
Butomus umbellatus . 964
Buxus sempervirens 896

Cakile maritima . . 110
Calamagrostis
epigeios . . . 1182
lanceolata . 1183
stricta . . . 1184
Calamintha acinos 787
clinopodium . 789
officinalis 788
Callitriche aquatica . 899
autumnalis . 899
hamulata . 899
pedunculata . 899
platycarpa . 899
Caltha palustris . . 24
Caluna vulgaris. 634
Calystegia sepium 684
soldanella . . 685
Camelina *fœtida* . 93
sativa 93
Campanula glomerata 614
Campanula hederacea 621
hybrida . . 622
latifolia 616
patula . . 619
rapunculoides . 617
rapunculus 618
rotundifolia 620
trachelium . . 615
Capsella bursa-pastoris 101
Cardamine amara 63
bulbifera . 67
hirsuta . . 66
impatiens. 65
pratensis . . 64
Carduus acanthoides . 557
acaulis . . 566
arvensis 561
crispus . 557
eriophorus . 562
heterophyllus . 563
lanceolatus 559
marianus . . 555
nutans . . 556
palustris . 560
pratensis . . 565
pycnocephalus 558
tenuifolia . . 558
tuberosus. . 564
Carex acuta . . 1123
alpina . . . 1124
ampullacea . . 1148
arenaria . . 1118
atrata . . 1126
axillaris. . . . 1114
bœnninghauseniana 1114
buxbaumii . 1125
cæspitosa. . . 1122
canescens 1112
canescens. . 1225
capillaris . . 1141
clandestina . 1127
collina . . 1130
curta . . . 1112
davalliana 1104
digitata . 1128
dioica . . . 1104
distans 1138
divisa . 1119
elongata . . 1110
extensa . . 1136
filiformis . . . 1133
flava . . . 1137
gibsoni . . 1123
glauca . 1143
grahami . . 1121
hirta. . . . 1134
humilis . 1127
incurva . . 1120
irrigua . 1142
lagopina . . 1109
leporina . . 1108
leporina . . . 1109
limosa . . . 1142
montana . . . 1130
muricata . . 1117
ovalis . . . 1108
pallescens . . 1135
paludosa . . 1150
panicea . 1140
paniculata . 1115
pauciflora . 1107
pendula . . 1147
Carex pilulifera . . 1131
præcox . . . 1129
pseudocyperus. . 1146
pulicaris . . 1105
pulla . . 1121
punctata . 1139
remota . 1113
rupestris . . 1106
saxatilis 1121
stellulata . 1111
strigosa . . 1145
sylvatica . . 1144
tomentosa . 1132
vahlii . 1124
vesicaria . . 1149
vulgaris . . 1122
vulpina . . 1116
Carlina vulgaris . . 568
Carpinus betulus . 911
Carum bulbocastanum 411
carvi . . 410
petroselinum 407
segetum . 408
verticillatum . . 409
Catabrosa aquatica . 1243
Caucalis anthriscus . 445
daucoides 447
infesta . . 446
latifolia . . . 448
nodosa . . 444
Centaurea aspera . 572
calcitrapa 573
cyanus 571
isnardi 572
nigra . 569
nigrescens 569
scabiosa . 570
solstitialis . 574
Centranthus ruber . 481
Centunculus minimus. 659
Cephalanthera ensifolia 982
grandiflora . 981
rubra . . . 983
Cerastium alpinum 160
arvense . . 159
glomeratum . . 158
quaternellum . . 156
trigynum. 161
vulgatum. 158
Ceratophyllum
demersum . . 898
submersum . . 898
Ceterach officinarum . 1297
Chærophyllum
anthriscus 443
sylvestre . . . 442
temulum . . 441
Chamagrostis minima 1172
Cheiranthus cheiri 50
Chelidonium majus . 42
Chenopodium album . 842
bonus-henricus . 848
botryoides . . 844
ficifolium . 842
glaucum . . 843
hybridum . . 847
murale . . 846
olidum . . . 840
polyspermum . . 841
rubrum . 844
urbicum . 845
vulvaria . . 840

FIG.

Cherleria sedoides 148
Chlora perfoliata 679
Chrysanthemum
 leucanthemum 520
 parthenium 522
 segetum 521
Chrysosplenium
 alternifolium 387
 oppositifolium 386
Cicendia filiformis 671
 pusilla 672
Cichorium intybus 603
Cicuta virosa 4[illegible]
Circæa alpina 352
 lutetiana 351
Cladium mariscus 1080
Claytonia perfoliata 172
Clematis vitalba 1
Cnicus acaulis 566
 arvensis 561
 eriophorus 562
 heterophyllus 563
 lanceolatus 559
 palustris 560
 pratensis 565
 tuberosus 564
Cochlearea armoracia 84
 officinalis 85
 polymorpha 85
Colchicum autumnale 1054
Comarum palustre 319
Conium maculatum 450
Conopodium
 denudatum 440
Convallaria majalis 1028
Convolvulus arvensis 683
 sepium 684
 soldanella 685
Corallorhiza innata 978
Coriandrum sativum 453
Cornus sanguinea 457
 suecica 456
Corrigiola littoralis 831
Corydalis claviculata 47
 lutea 46
Corylus avellana 912
Corynephorus
 canescens 1187
Cotoneaster vulgaris 338
Cotyledon umbilicus 358
Crambe maritima 111
Cratægus oxyacantha 337
Crepis biennis 596
 fœtida 594
 hieracioides 597
 paludosa 598
 succisaefolia 597
 taraxacifolia 593
 virens 595
Crithmum maritimum 430
Crocus nudiflorus 1018
 vernus 1017
Crytogramme
 crispa 1278
 leptophylla 1277
Cuscuta epilinum 687
 epithymum 688
 europæa 686
 trifolii 688
Cyclamen europæum 650
 hederæfolium 650
Cynodon dactylon 1196

FIG.

Cynoglossum
 montanum 709
 officinale 708
 sylvaticum 709
Cynosurus cristatus 1224
 echinatus 1225
Cyperus fuscus 1078
 longus 1077
Cypripedium calceolus 1011
Cystopteris *dentata* 1301
 fragilis 1301
 montana 1302
Cytisus scoparius 230

Dabeocia polifolia 632
Dactylis glomerata 1223
Damasonium stellatum 969
Daphne laureola 878
 mezereum 877
Datura stramonium 710
Daucus carota 449
Delphinium ajacis 29
Dentaria bulbifera 67
Dianthus armeria 128
 cæsius 130
 deltoides 129
 prolifer 127
Digitalis purpurea 746
Digitaria humifusa 1154
 sanguinalis 1153
Digraphis arundinacea 1162
Diotis maritima 531
Diplotaxis muralis 76
 tenuifolia 75
Dipsacus pilosus 490
 sylvestris 489
Doronicum
 pardalianches 550
 plantagineum 551
Draba aizoides 88
 hirta 89
 incana 90
 muralis 91
 rupestris 89
 verna 92
Drosera anglica 391
 longifolia 390
 rotundifolia 389
Dryas octopetala 302

Echinochola crus-galli 1158
Echium *plantagineum* 690
 violaceum 690
 vulgare 689
Elatine hexandra 175
 hydropiper 176
Eleocharis acicularis 1085
 cæspitosa 1090
 multicaulis 1088
 palustris 1087
 parvula 1086
 pauciflora 1089
Elodea canadensis 973
Elymus arenarius 1200
Empetrum nigrum 897
Epilobium alpinum 348
 alsinæfolium 347
 anagallidifolium 348
 angustifolium 340
 hirsutum 341
 montanum 343
 obscurum 345
 palustre 346

FIG.

Epilobium parviflorum 342
 roseum 344
 tetragonum 345
Epipactis *ensifolia* 983
 grandiflora 981
 latifolia 979
 media 979
 ovalis 979
 palustris 980
 rubra 983
Epipogium aphyllum 987
 gmelini 987
Equisetum arvense 1260
 hyemale 1265
 limosum 1263
 maximum 1259
 moorei 1266
 palustre 1264
 pratense 1262
 ramosum 1266
 sylvaticum 1261
 telmateia 1259
 trachyodon 1266
 umbrosum 1262
 variegatum 1267
Erica carnea 638
 ciliaris 637
 cinerea 635
 mediterranea 638
 tetralix 636
 vagans 639
 vulgaris 634
Erigeron acris 497
 alpinus 498
 canadensis 499
Eriocaulon
 septangulare 1076
Eriophorum alpinum 1100
 polystachyum 1102
 vaginatum 1101
Erodium cicutarium 212
 maritimum 214
 moschatum 213
Erophila verna 92
Eryngium campestre 399
 maritimum 398
Erysimum
 cheiranthoides 73
 orientale 74
Erythræa centaurium 673
Eupatorium
 cannabinum 494
Euphorbia
 amygdaloides 893
 esula 892
 exigua 888
 helioscopia 883
 hiberna 885
 lathyris 889
 palustris 886
 paralias 891
 peplis 882
 peplus 887
 pilosa 886
 platyphyllos 884
 portlandica 890
 segetalis 890
 stricta 884
Euphrasia *odontites* 765
 officinalis 766
 viscosa 764
Evonymus europæus 222

	FIG.
Fagus sylvatica	913
Fedia auricula	487
carinata	486
dentata	488
olitoria	485
Festuca *ambigua*	1221
bromoides	1221
elatior	1219
gigantea	1217
myurus	1221
ovina	1218
sciuroides	1221
sylvatica	1220
uniglumis	1222
Filago *apiculata*	502
gallica	504
germanica	502
minima	503
spathulata	502
Fœniculum *officinale*	425
vulgare	425
Fragaria vesca	310
Frankenia lævis	126
Fraxinus excelsior	667
Fritillaria meleagris	1032
Fumaria officinalis	45
Gagea lutea	1035
Galanthus nivalis	1021
Galeobdolon luteum	811
Galeopsis *dubia*	803
ladanum	802
ochroleuca	803
tetrahit	804
Galium *anglicum*	474
aparine	476
boreale	475
cruciata	468
mollugo	473
palustre	470
parisiense	474
saxatile	472
tricorne	477
uliginosum	471
verum	469
Gastridium	
lendigerum	1180
Genista anglica	229
pilosa	228
tinctoria	227
Gentiana amarella	677
campestris	678
nivalis	676
pneumonanthe	674
verna	675
Geranium columbinum	211
dissectum	210
lucidum	206
molle	207
phœum	201
pratense	203
pusillum	208
pyrenaicum	204
robertianum	205
rotundifolium	209
sanguineum	200
sylvaticum	202
Geum rivale	304
urbanum	303
Githago segetum	141
Gladiolus communis	1014
illyricus	1014
Glaucium luteum	44
Glaux maritima	656
Glyceria aquatica	1228
distans	1231
fluitans	1229
loliacea	1234
maritima	1230
plicata	1229
procumbens	1232
rigida	1233
Gnaphalium	
luteo-album	505
supinum	507
sylvaticum	506
uliginosum	508
Goodyera repens	991
Grammitis	
leptophylla	1277
Gymnadenia albida	1004
conopsea	1002
Habenaria albida	1004
bifolia	1003
viridis	1005
Haloscias scoticum	427
Hedera helix	454
Helianthemum canum	117
guttatum	116
polifolium	119
vulgare	118
Helleborus foetidus	27
viridis	26
Helminthia echioides	577
Heracleum	
sphondylium	436
Herminium monorchis	1007
Herniaria *ciliata*	832
glabra	832
Hesperis matronalis	68
Hieracium alpinum	600
boreale	604
cerinthoides	602
murorum	601
pilosella	599
prenanthoides	605
sabaudum	604
umbellatum	603
Hierochloe borealis	1159
Hippocrepis comosa	274
Hippophae	
rhamnoides	879
Hippuris vulgaris	394
Holcus lanatus	1194
mollis	1195
Holosteum	
umbellatum	157
Honckenya peploides	152
Hordeum maritimum	1204
murinum	1203
pratense	1202
sylvaticum	1201
Hottonia palustris	646
Humulus lupulus	904
Hutchinsia petræa	100
Hydrocharis	
morsus-ranæ	974
Hydrocotyle vulgaris	395
Hymenophyllum	
tunbridgense	1305
unilaterale	1306
wilsoni	1306
Hyoscyamus niger	711
Hypericum	
androsæmum	178
calycinum	177
dubium	180
elodes	187
hirsutum	185
humifusum	182
linariifolium	183
montanum	186
perforatum	179
pulchrum	184
quadrangulum	181
Hypochœris glabra	582
maculata	584
radicata	583
Iberis amara	99
Ilex aquifolium	221
Illecebrum	
verticillatum	833
Impatiens fulva	218
noli-me-tangere	217
Inula conyza	514
crithmoides	513
dysenterica	515
helenium	511
pulicaria	516
salicina	512
Iris fœtidissima	1013
pseudacorus	1012
Isatis tinctoria	109
Isnardia palustris	350
Isoetes duriæi	1252
lacustris	1251
Isolepis fluitans	1091
holoschænus	1094
saviana	1093
savii	1093
setacea	1092
Jasione montana	611
Juncus acutus	1067
articulatus	1059
balticus	1058
biglumis	1070
bufonius	1063
capitatus	1065
castaneus	1069
communis	1055
compressus	1061
diffusus	1056
filiformis	1057
gerardi	1061
glaucus	1056
maritimus	1066
obtusiflorus	1060
pygmæus	1064
squarrosus	1062
trifidus	1068
Juniperus communis	934
Knappia agrostidea	1172
Knautia arvensis	493
Kobresia caricina	1103
Kœleria cristata	1248
Koniga maritimum	87
Lactuca muralis	585
saligna	587
scariola	586
Lagurus ovatus	1173
Lamium album	809
amplexicaule	807

FIG.

Lamium album
  galeobdolon 811
  maculatum 810
  purpureum 808
Lapsana communis 608
  pusilla 607
Lastrea cristata 1283
  dilatata 1284
  filix-mas 1282
  oreopteris 1281
  rigida 1285
  spinulosa 1284
  thelypteris 1280
Lathræa squamaria 722
Lathyrus aphaca 287
  hirsutus 288
  macrorrhizus 294
  maritimus 293
  niger 295
  nissolia 286
  palustris 292
  pratensis 289
  sylvestris 291
  tuberosus 290
Lavatera arborea 193
Leersia oryzoides 1151
Lemna arrhiza 947
  gibba 945
  minor 944
  polyrrhiza 946
  trisulca 943
Leontodon autumnalis 580
  hirtus 581
  hispidus 579
  taraxacum 592
Leonurus cardiaca 806
Lepidium campestre 102
  draba 104
  latifolium 105
  ruderale 106
  smithii 103
Lepigonum rubra 169
Lepturus incurvatus 1198
Leucoium æstivum 1022
Ligusticum scoticum 427
Ligustrum vulgare 668
Limnanthemum
  nymphæoides 681
Limosella aquatica 744
Linaria cymbalaria 736
  elatine 738
  minor 735
  pelisseriana 733
  repens 732
  spuria 737
  supina 734
  vulgaris 731
Linnæa borealis 466
Linum angustifolium 190
  catharticum 191
  perenne 189
  usitatissimum 188
Liparis loeselii 977
Listera cordata 985
  nidus-avis 986
  ovata 984
Litho-spermum
  arvense 693
  officinale 694
  purpureo-cæruleum 695
Littorella lacustris 830
Lloydia serotina 1034

Lobelia dortmanna 609
  urens 610
Loiseleuria
  procumbens 631
Lolium perenne 1207
  temulentum 1208
Lomaria spicant 1298
Lonicera
  caprifolium 464
  periclymenum 463
  xylosteum 465
Lotus angustissimus 265
  corniculatus 264
Ludwigia palustris 350
Luzula arcuata 1073
  campestris 1074
  multiflora 1074
  pilosa 1071
  spicata 1075
  sylvatica 1072
Lychnis alpina 144
  diurna 140
  flos-cuculi 142
  githago 141
  vespertina 139
  viscaria 143
Lycopodium alpinum 1255
  annotinum 1254
  clavatum 1253
  inundatum 1257
  selaginoides 1258
  selago 1256
Lycopsis arvensis 703
Lycopus europæus 776
Lynosiris vulgaris 496
Lysimachia nemorum 654
  nummularia 653
  thyrsiflora 652
  vulgaris 651
Lythrum
  hyssopifolium 354
  salicaria 353

Maianthemum bifolium 1029
Malaxis paludosa 976
Mallachium aquaticum 162
Malva moschata 196
  rotundifolia 194
  sylvestris 195
Marrubium vulgare 796
Matricaria chamomilla 524
  inodora 523
Matthiola incana 48
  sinuata 49
Matricaria parthenium 522
Meconopsis cambrica 41
Medicago denticulata 236
  falcata 233
  lupulina 235
  maculata 237
  minima 238
  sativa 234
  sylvestris 233
Melampyrum arvense 771
  cristatum 770
  pratense 772
  sylvaticum 773
Melica nutans 1245
  uniflora 1246
Melilotus alba 241
  arvensis 240

Melilotus officinalis 239
  vulgaris 241
Melittis
  melissophyllum 795
Mentha aquatica 781
  arvensis 783
  piperita 780
  pulegium 784
  rotundifolia 778
  sativa 782
  silvestris 777
  viridis 779
Menyanthes trifoliata 680
Menziesia cærulea 633
  polifolia 632
Mercurialis annua 895
  perennis 894
Mertensia maritima 692
Mespilus germanica 339
Meum athamanticum 429
Microcala filiformis 671
Milium effusum 1152
Mimulus luteus 743
Mœnchia erecta 156
Molinia cærulia 1244
Moneses grandiflora 640
Monotropa hypopitys 645
Montia fontana 173
Mulgedium alpinum 591
Muscari racemosum 1042
Myosotis arvensis 698
  collina 699
  palustris 696
  sylvatica 697
  versicolor 700
Myosurus minimus 8
Myrica gale 907
Myriophyllum
  spicatum 392
  verticillatum 393
Myrrhis odorata 439

Naias flexilis 950
Narcissus biflorus 1020
  pseudonarcissus 1019
Nardus stricta 1199
Narthecium
  ossifragum 1052
Nasturtium
  amphibium 55
  officinale 52
  palustre 54
  sylvestre 53
  terestre 54
Neottia nidus-avis 986
Nepeta cataria 791
  glechoma 790
Nephrodium cristatum 1283
  filix-mas 1282
  oreopteris 1281
  rigidum 1285
  spinulosum 1284
  thelypteris 1280
Nuphar lutea 35
Nymphæa alba 34

Œnanthe crocata 422
  fistulosa 420
  phellandrium 423
  pimpinelloides 421
Œnothera biennis 349
Onobrychis sativa 275

FIG.
Ononis arvensis . 231
campestris 231
reclinata . 232
Onopordon acanthium 567
Ophioglossum
vulgatum. 1269
Ophrys apifera 1008
arachnites 1008
aranifera . 1009
muscifera. 1010
Orchis conopsea . 1002
hircina 1000
intacta 995
latifolia 999
laxiflora . 997
maculata . 998
mascula 996
militaris . 998
morio 992
purpuria. 993
pyramidalis 1001
simia 993
ustulata 994
Origanum vulgare 786
Ornithogalum nutans . 1037
pyrenaicum 1036
umbellatum 1035
Ornithopus
ebracteatus 272
perpusillus 273
Orobanche cærulea 720
caryophyllacea 716
elatior 718
major 715
minor 719
ramosa 721
rapum 715
rubra 717
Osmunda regalis 1271
Oxalis acetosella. 215
corniculata 216
Oxycoccus palustris 626
Oxyria reniformis 865
Oxytropis campestris. 270
halleri 271
uralensis . 271

Pæonia officinalis 32
corallina. 32
Panicum crus-galli 1158
glabrum . 1154
glaucum . 1156
sanguinale 1153
verticillatum 1155
viride 1157
Papaver argemone 40
dubium . 38
hybridum 39
lecoqii 38
rhœas 37
somniferum 36
Parietaria *diffusa* 903
officinalis. 903
Paris quadrifolia. 1024
Parnassia palustris 388
Pastinaca sativa. 435
Pedicularis palustris . 768
sylvatica . 769
Peplis portula 355
Petasites vulgaris 538
Peucedanum officinale 432
ostruthium 434

FIG.
Peucedanum palustre. 433
sativum 435
Phalaris *arundinacea*. 1162
canariensis 1161
Phleum alpinum. 1164
arenarium 1167
asperum . 1166
bœhmeri . 1165
pratense . 1163
Phragmites communis 1250
Phyllodoce caerulea 633
Physospermum
cornubiense 451
Phyteuma orbiculare . 612
spicatum . 613
Picris hieracioides 578
Pilularia globulifera . 1268
Pimpinella magna 415
saxifraga . 414
Pinguicula alpina 662
lusitanica. 663
vulgaris . 661
Pinus sylvestris . 933
Plantago coronopus 829
lanceolata 827
major 825
maritima . 828
media 826
Poa alpina . 1241
annua 1235
aquatica . 1228
balfourii. 1239
borreri 1231
bulbosa 1242
compressa 1236
distans 1231
fluitans 1229
laxa . 1240
loliacea 1234
maritima . 1230
minor 1240
nemoralis. 1239
parnellii . 1239
pratensis. 1237
procumbens 1232
rigida 1233
trivialis 1238
Polemonium cæruleum 68[illegible]
Polycarpon
tetraphyllum . 171
Polygala *calcarea* 125
vulgaris 125
Polygonatum
multiflorum 1026
officinale . 1027
verticillatum 1025
Polygonum amphibium 872
aviculare 866
bistorta . 871
convolvulus 868
dumetorum 869
flexile 1274
hydropiper 875
lapathifolium . 874
maritimum 867
minus 876
persicaria. 873
viviparum 870
Polypodium alpestre. 1274
dryopteris 1275
phegopteris 1273
vulgare 1272

FIG.
Polypogon littoralis 1175
monspeliensis . 1174
Polystichum aculeatum. 1279
lonchitis. 1278
Populus alba 930
nigra 932
tremula 931
Potamogeton
acutifolius 961
crispus 958
densus 959
filiformis 963
heterophyllus . 954
longifolius 955
lucens 955
natans 953
obtusifolius 960
pectinatus 963
perfoliatus 957
polygonifolius. 953
prælongus 956
pusillus 962
trichoides 962
Potentilla anserina 317
argentea . 314
comarum . 319
fragariastrum . 311
fruticosa . 316
procumbens 320
reptans 312
rupestris . 318
tormentilla 313
verna 315
Poterium *officinale* 324
sanguisorba 325
Primula farinosa. 649
veris 648
vulgaris 647
Prunella vulgaris 792
Prunus cerasus 297
communis 296
padus 298
Psamma arenaria 1181
Pteris aquilina 1299
Pulicaria dysenterica 515
vulgaris. 516
Pulmonaria officinalis. 691
Pyrola media 642
minor 643
rotundifolia 641
secunda 644
uniflora 640
Pyrus aria . 334
aucuparia. 336
communis 332
germanica 339
malus 333
torminalis 335

Quercus robur 914

Radiola millegrana 192
Ranunculus acris 17
aquatilis . 9
arvensis . 23
auricomus 16
bulbosus . 20
chaerophyllus . 19
ficaria 14
flammula . 12
hederaceus 10

	FIG.
Ranunculus *hirsutus*	21
lingua	11
ophioglossifolius	13
parviflorus	22
philonotis	21
repens	18
sceleratus	15
Raphanus raphanistrum	112
Reseda alba	115
fruticulosa	115
lutea	114
luteola	113
Rhamnus catharticus	223
frangula	224
Rhinanthus crista-galli	767
Rhynchospora alba	1082
fusca	1081
Ribes alpinum	371
grossularia	369
nigrum	372
rubrum	370
Rœmeria hybrida	43
Romulea columnæ	1016
Rosa arvensis	331
caesia	330
canina	330
micrantha	329
pimpinellifolia	327
rubella	327
rubiginosa	329
sepium	329
spinonissima	327
tomentosa	328
villosa	328
Rubia peregrina	457
Rubus cæsius	307
chamæmorus	309
fruticosus	306
idæus	305
saxatilis	308
Rumex acetosa	863
acetosella	864
aquaticus	855
conglomeratus	859
crispus	856
hydrolapathum	858
maritimus	862
obtusifolius	857
pulcher	861
sanguineus	860
Ruppia maritima	952
Ruscus aculeatus	1031
Sagina *apetala*	145
ciliata	145
linnæi	146
nodosa	147
procumbens	145
saxatilis	146
subulata	146
Sagittaria sagittifolia	965
Salicornia herbacea	836
radicans	836
Salix alba	917
amygdalina	918
aurita	923
caprea	921
fragilis	916
herbacea	929
lanata	926
lapponum	925
Salix myrsinites	927
pentandra	915
phylicifolia	923
purpurea	919
repens	924
reticulata	928
triandra	918
viminalis	920
Salsola kali	839
Salvia pratensis	774
verbenaca	775
Sambucus ebulus	460
nigra	459
Samolus valerandi	660
Sanguisorba officinalis	334
Sanicula europæa	396
Saponaria officinalis	131
Sarothamnus scoparius	230
Saussurea alpina	554
Saxifraga *affinis*	376
aizoides	374
cæspitosa	377
cernua	379
geum	385
granulata	378
hirculus	375
hirta	376
hypnoides	376
nivalis	382
oppositifolia	373
rivularis	380
stellaris	383
tridactylites	381
umbrosa	384
Scabiosa arvensis	493
columbaria	492
succisa	491
Scandix pecten	438
Scheuchzeria palustris	970
Schœnus nigricans	1079
Scilla autumnalis	1040
nutans	1041
verna	1039
Scirpus acicularis	1085
cæspitosus	1090
fluitans	1091
holoschœnus	1094
lacustris	1097
maritimus	1098
multicaulis	1088
palustris	1087
parvulus	1086
pauciflorus	1089
pungens	1095
riparius	1093
rothii	195
setaceus	1092
sylvaticus	1099
triqueter	1096
Scleranthus annuus	834
perennis	835
Sclerochloa distans	1231
loliacea	1234
maritima	1230
procumbens	1232
rigida	1233
Scolopendrium vulgare	1296
Scrophularia aquatica	740
nodosa	739
scorodonia	741
vernalis	742
Scutellaria galericulata	793
minor	794
Sedum acre	365
album	363
anglicum	361
dasyphyllum	362
fabarium	360
rhodiola	359
rupestre	367
sexangulare	366
telephium	360
villosum	364
Selaginella selaginoides	1258
Sempervivum tectorum	368
Senebiera coronopus	107
didyma	108
Senecio aquaticus	543
campestris	549
erucifolius	545
jacobæa	544
paludosus	546
palustris	548
saracenicus	547
squalidus	542
sylvaticus	541
tenuifolius	545
viscosus	540
vulgaris	539
Serrafalcus arvensis	1216
Serratula tinctoria	553
Seseli libanotis	426
Sesleria cærulea	1249
Setaria glauca	1156
verticillatum	1155
viridis	1157
Sherardia arvensis	480
Sibbaldia procumbens	320
Sibthorpia europæa	745
Silaus pratensis	428
Silene acaulis	132
anglica	136
conica	137
gallica	136
inflata	133
noctiflora	138
nutans	135
otites	134
Silybum marianus	555
Simethis bicolor	1051
Sinapis alba	80
arvensis	81
incana	83
nigra	82
Sison amomum	404
Sisymbrium *alliaria*	72
irio	70
officinale	69
sophia	71
thaliana	60
Sisyrinchium *anceps*	1015
bermudiana	1015
Sium angustifolium	413
latifolium	412
Smilacina bifolia	1029
Smyrnium olusatrum	452
Solanum dulcamara	712
nigrum	713
Solidago virga-aurea	500
Sonchus alpinus	591
arvenis	588
oleraceus	590

FIG.
Sonchus palustris 589
Sparganium minimum 940
  natans 940
  ramosum 938
  simplex 939
Spartina stricta 1197
Specularia hybrida 622
Spergula arvensis 170
Spergularia rubra 169
Spiræa filipendula 301
  salicifolia 299
  ulmaria 300
Spiranthes æstivalis 989
  autumnalis 988
  cernua 990
  gemmipara 990
  romanzoviana 990
Stachys arvensis 801
  betonica 797
  germanica 798
  palustris 800
  sylvatica 799
Statice auriculifolia 821
  bellidifolia 823
  binervosa 831
  dodartii 821
  limonium 820
  occidentalis 821
  reticulata 822
Stellaria aquatica 162
  glauca 167
  graminea 166
  holostea 168
  media 164
  nemorum 163
  uliginosa 165
Stratiotes aloides 975
Suæda fruticosa 837
  maritima 838
Subularia aquatica 94
Symphytum officinale 704
  tuberosum 705

Tamarix *anglica* 174
  gallica 174
Tamus communis 1023
Tanacetum vulgare 532
Taraxacum dens-leonis 592
  officinale 592
Taxus baccata 935
Teesdalia nudicaulis 98
Teucrium botrys 814
  chamædrys 815
  scordium 813
  scorodonia 812
Thalictrum alpinum 2
  flavum 4
  flexuosum 3
  kochii 3
  minus 3
  saxatile 3
Thesium *humifusum* 880
  linophyllum 880
Thlaspi alpestre 97
  arvense 95
  perfoliatum 96
Thrincia hirtus 581
Thymus *chamædrys* 785
  serpyllum 785

FIG.
Tilia *parvifolia* 199
  europæa 199
Tillæa muscosa 357
Tofieldia palustris 1053
Tordylium maximum 437
Torilis anthriscus 446
  infesta 447
  nodosa 445
Tragopogon *minor* 575
  porrifolius 576
  pratensis 575
Trichomanes
  radicans 1304
Trichonema
  bulbocodium 1016
  columnæ 1016
Trientalis europæa 655
Trifolium arvense 244
  bocconi 251
  filiforme 263
  fragiferum 258
  glomeratum 254
  hybridum 260
  incarnatum 243
  maritimum 249
  medium 248
  minus 262
  ochroleucum 246
  pratense 247
  procumbens 261
  repens 259
  resupinatum 256
  scabrum 252
  stellatum 245
  striatum 250
  strictum 253
  subterraneum 257
  suffocatum 255
Triglochin maritimum 972
  palustre 971
Trigonella
  ornithopodioides 242
Trinia vulgaris 405
Triodia decumbens 1247
Trisetum flavescens 1192
Triticum caninum 1206
  repens 1205
Trollius europæus 25
Tulipa sylvestris 1033
Turrites glabra 56
Tussilago farfara 537
  petasites 538
Typha angustifolia 937
  latifolia 936

Ulex europæus 225
  nanus 226
Ulmus campestris 906
  campestris 905
  montana 905
  suberosa 906
Urtica dioica 902
  pilulifera 901
  urens 900
Utricularia intermedia 666
  minor 665
  vulgaris 664

Vaccinium myrtillus 623

FIG.
Vaccinium oxycoccos 626
  uliginosum 624
  vitis-idæa 625
Valeriana dioica 482
  officinalis 483
  pyrenaica 484
Valerianella auricula 487
  carinata 486
  dentata 488
  olitoria 485
Verbascum blattaria 724
  lychnitis 727
  nigrum 726
  pulverulentum 728
  thapsus 723
  virgatum 725
Verbena officinalis 819
Veronica agrestis 758
  alpina 749
  anagallis 752
  arvensis 760
  beccabunga 753
  buxbaumii 759
  chamædrys 756
  hederæfolia 757
  montana 755
  officinalis 751
  saxatilis 748
  scutellata 754
  serpyllifolia 750
  spicata 747
  triphyllos 762
  verna 761
Viburnum lantana 461
  opulus 462
Vicia bithynica 285
  cracca 278
  hirsuta 276
  lathyroides 284
  lutea 282
  orobus 280
  sativa 283
  sepium 281
  sylvatica 279
  tetrasperma 277
Villarsia
  nymphæoides 681
Vinca major 669
  minor 670
Viola canina 123
  hirta 122
  odorata 121
  palustris 120
  stagnina 123
  sylvatica 123
  tricolor 124
Viscum album 455

Wahlenbergia
  hederacia 621
Woodsia ilvensis 1303

Xanthium strumarium 517

Zannichellia palustris 951
Zostera marina 948
  nana 949

LIST OF WORKS

ON

BOTANY, ENTOMOLOGY, CONCHOLOGY,

TRAVELS, TOPOGRAPHY,

ANTIQUITY, AND MISCELLANEOUS

LITERATURE AND SCIENCE.

PUBLISHED BY

L. REEVE AND CO.,

5, HENRIETTA STREET, COVENT GARDEN, W.C.

NEW SERIES OF POPULAR NATURAL HISTORY FOR BEGINNERS AND AMATEURS.

British Insects; a Familiar Description of the Form, Structure, Habits, and Transformations of Insects. By E. F. Staveley. Crown 8vo, 16 Coloured Plates, and numerous Wood Engravings, 14*s*.

British Butterflies and Moths; an Introduction to the Study of our Native Lepidoptera. By H. T. Stainton. Crown 8vo, 16 Coloured Plates, and Wood Engravings, 10*s*. 6*d*.

British Beetles; an Introduction to the Study of our indigenous Coleoptera. By E. C. Rye. Crown 8vo, 16 Coloured Plates, and 11 Wood Engravings, 10*s*. 6*d*.

British Bees; an Introduction to the Study of the Natural History and Economy of the Bees indigenous to the British Isles. By W. E. Shuckard. Crown 8vo, 16 Coloured Plates, and Woodcuts, 10*s*. 6*d*.

British Spiders; an Introduction to the Study of the Araneidæ found in Great Britain and Ireland. By E. F. Staveley. Crown 8vo, 16 Coloured Plates, and 44 Wood Engravings, 10*s*. 6*d*.

British Grasses; an Introduction to the Study of the Grasses found in the British Isles. By M. Plues. Crown 8vo, 16 Coloured Plates, and 100 Wood Engravings, 10*s*. 6*d*.

British Ferns; an Introduction to the Study of the Ferns, Lycopods, and Equiseta indigenous to the British Isles. With Chapters on the Structure, Propagation, Cultivation, Diseases, Uses, Preservation, and Distribution of Ferns. By M. Plues. Crown 8vo, 16 Coloured Plates, and 55 Wood Engravings, 10*s*. 6*d*.

British Seaweeds; an Introduction to the Study of the Marine ALGÆ of Great Britain, Ireland, and the Channel Islands. By S. O. GRAY. Crown 8vo, 16 Coloured Plates, 10*s*. 6*d*.

BOTANY.

The Natural History of Plants. By H. BAILLON, President of the Linnæan Society of Paris, Professor of Medical Natural History and Director of the Botanical Garden of the Faculty of Medicine of Paris. Super-royal 8vo. Vols I. to VII., with 3200 Wood Engravings, 25*s*. each.

Handbook of the British Flora; a Description of the Flowering Plants and Ferns indigenous to, or naturalized in, the British Isles. For the use of Beginners and Amateurs. By GEORGE BENTHAM, F.R.S. 4th Edition, revised, Crown 8vo, 12*s*.

Illustrations of the British Flora; a Series of Wood Engravings, with Dissections, of British Plants, from Drawings by W. H. FITCH, F.L.S., and W. G. SMITH, F.L.S., forming an Illustrated Companion to BENTHAM'S "Handbook," and other British Floras. 1306 Wood Engravings, 12*s*.

Domestic Botany; an Exposition of the Structure and Classification of Plants, and of their uses for Food, Clothing, Medicine, and Manufacturing Purposes. By JOHN SMITH, A.L.S., ex-Curator of the Royal Gardens, Kew. Crown 8vo, 16 Coloured Plates and Wood Engravings, 16*s*.

British Wild Flowers, Familiarly Described in the Four Seasons. By THOMAS MOORE, F.L.S. 24 Coloured Plates, 16*s*.

The Narcissus, its History and Culture, with Coloured Figures of all known Species and Principal Varieties. By F. W. BURBIDGE, and a Review of the Classification by J. G. BAKER, F.L.S. Super-royal 8vo, 48 Coloured Plates, 32*s*.

The Botanical Magazine; Figures and Descriptions of New and Rare Plants of interest to the Botanical Student, and suitable for the Garden, Stove, or Greenhouse. By Sir J. D. HOOKER, K.C.S.I., C.B., F.R.S., Director of the Royal Gardens, Kew. Royal 8vo. Third Series, Vols. I. to XXXVIII., each 42*s*. Published Monthly, with 6 Plates, 3*s*. 6*d*., coloured. Annual Subscription, 42*s*.

RE-ISSUE of the THIRD SERIES in Monthly Vols., 42*s*. each; to Subscribers for the entire Series, 36*s*. each.

The Floral Magazine; New Series, Enlarged to Royal 4to. Figures and Descriptions of the choicest New Flowers for the Garden, Stove, or Conservatory. Complete in Ten Vols., in handsome cloth, gilt edges, 42*s*. each.

FIRST SERIES complete in Ten Vols., with 560 beautifully-coloured Plates, £18 7*s*. 6*d*.

Wild Flowers of the Undercliff, Isle of Wight. By CHARLOTTE O'BRIEN and C. PARKINSON. Crown 8vo, 8 Coloured Plates, 7*s*. 6*d*.

The Young Collector's Handybook of Botany. By the Rev. H. P. DUNSTER, M.A. 66 Wood Engravings, 3*s*. 6*d*.

Laws of Botanical Nomenclature adopted by the International Botanical Congress, with an Historical Introduction and a Commentary. By ALPHONSE DE CANDOLLE. 2*s*. 6*d*.

Contributions to the Flora of Mentone, and to a Winter Flora of the Riviera, including the Coast from Marseilles to Genoa. By J. TRAHERNE MOGGRIDGE, F.L.S. Royal 8vo. Complete in One Vol., with 99 Coloured Plates, 63*s*.

Flora Vitiensis; a Description of the Plants of the Viti or Fiji Islands, with an Account of their History, Uses, and Properties. By Dr. BERTHOLD SEEMANN, F.L.S. Royal 4to, Coloured Plates. Part X., 25*s*.

Flora of Mauritius and the Seychelles; a Description of the Flowering Plants and Ferns of those Islands. By J. G. Baker, F.L.S. 24*s*. Published under the authority of the Colonial Government of Mauritius.

Flora of British India. By Sir J. D. Hooker, K.C.S.I., C.B., F.R.S., &c.; assisted by various Botanists. Parts I. to X., 10*s*. 6*d*. each. Vols. I. to III., cloth, 32*s*. each. Published under the authority of the Secretary of State for India in Council.

Flora of Tropical Africa. By Daniel Oliver, F.R.S., F.L.S. Vols. I. to III., 20*s*. each. Published under the authority of the First Commissioner of Her Majesty's Works.

Handbook of the New Zealand Flora; a Systematic Description of the Native Plants of New Zealand, and the Chatham, Kermadec's, Lord Auckland's, Campbell's, and Mac quarrie's Islands. By Sir J. D. Hooker, K.C.S.I., F.R.S. Part II., CRYPTOGAMIA, 14*s*. Published under the auspices of the Government of that Colony.

Flora Australiensis; a Description of the Plants of the Australian Territory. By George Bentham, F.R.S., assisted by Ferdinand Mueller, F.R.S., Government Botanist, Melbourne, Victoria. Complete in Seven Vols., £7 4*s*. Vols. I. to VI., 20*s*. each; Vol. VII., 24*s*. Published under the auspices of the several Governments of Australia.

Flora of the British West Indian Islands. By Dr. Grisebach, F.L.S. 42*s*. Published under the auspices of the Secretary of State for the Colonies.

Flora Hongkongensis; a Description of the Flowering Plants and Ferns of the Island of Hongkong. By George Bentham, F.R.S. With a Map of the Island, and a Supplement by Dr. Hance. 18*s*. Published under the authority of Her Majesty's Secretary of State for the Colonies. The Supplement separately, 2*s*. 6*d*.

Flora Capensis; a Systematic Description of the Plants of the Cape Colony, Caffraria, and Port Natal. By WILLIAM H. HARVEY, M.D., F.R.S., Professor of Botany in the University of Dublin, and OTTO WILHEM SONDER, Ph.D. Vols. I. and II., 12*s*. each. Vol. III., 18*s*.

Flora of Hampshire, including the Isle of Wight, with localities of the less common species. By FREDERICK TOWNSEND, M.A., F.L.S. With Coloured Map and two Plates, 16*s*.

On the Flora of Australia: its Origin, Affinities, and Distribution; being an Introductory Essay to the "Flora of Tasmania." By Sir J. D. HOOKER, F.R.S. 10*s*.

Genera Plantarum, ad Exemplaria imprimis in Herbariis Kewensibus servata definita. By GEORGE BENTHAM, F.R.S., F.L.S., and Sir J. D. HOOKER, F.R.S., Director of the Royal Gardens, Kew. Complete in Three Vols., £8 5*s*.; or separately:—Vol. I.—Part I., Royal 8vo, 21*s*.; Part II., 14*s*; Part III., 15*s*.; or Vol. I. complete, 50*s*. Vol. II.—Part I., 24*s*.; Part II., 32*s*.; or Vol. II. complete, 56*s*. Vol. III.—Part I., 24*s*.; Part II., 36*s*.; or Vol. III. complete, 56*s*.

Illustrations of the Nueva Quinologia of Pavon, with Observations on the Barks described. By J. E. HOWARD, F.L.S. With 27 Coloured Plates. Imperial folio, half-morocco gilt edges, £6 6*s*.

The Quinology of the East Indian Plantations. By J. E. HOWARD, F.L.S. Complete in One Vol., folio. With 13 Coloured and 2 Plain Plates, and 2 Photo-prints, 84*s*. Parts II. and III., cloth, 63*s*.

Revision of the Natural Order Hederaceæ; being a reprint, with numerous additions and corrections, of a series of papers published in the "Journal of Botany, British and Foreign." By BERTHOLD SEEMANN, Ph.D., F.L.S. 7 Plates, 10*s*. 6*d*.

Icones Plantarum. Figures, with Brief Descriptive Characters and Remarks, of New and Rare Plants, selected from the Author's Herbarium. By Sir W. J. HOOKER, F.R.S. New Series, Vol. V. 100 Plates, 31*s*. 6*d*.

Orchids; and How to Grow them in India and other Tropical Climates. By SAMUEL JENNINGS, F.L.S., F.R.H.S., late Vice-President of the Agri-Horticultural Society of India. Royal 4to. Complete in One Vol., cloth, gilt edges, 63*s.*

A Second Century of Orchidaceous Plants, selected from the Subjects published in Curtis's "Botanical Magazine" since the issue of the "First Century." Edited by JAMES BATEMAN, Esq., F.R.S. Complete in One Vol., Royal 4to, 100 Coloured Plates, £5 5*s.*

Dedicated by Special Permission to H.R.H. the Princess of Wales.

Monograph of Odontoglossum, a Genus of the Vandeous Section of Orchidaceous Plants. By JAMES BATEMAN, Esq., F.R.S. Imperial folio, complete in Six Parts, each with 5 Coloured Plates, and occasional Wood Engravings, 21*s.*; or, in One Vol., cloth, £6 16*s.* 6*d.*

The Rhododendrons of Sikkim-Himalaya; being an Account, Botanical and Geographical, of the Rhododendrons recently discovered in the Mountains of Eastern Himalaya, by Sir J. D. Hooker, F.R.S. By Sir W. J. HOOKER, F.R.S. Folio, 30 Coloured Plates, £4 14*s.* 6*d.*

Outlines of Elementary Botany, as Introductory to Local Floras. By GEORGE BENTHAM, F.R.S., President of the Linnæan Society. New Edition, 1*s.*

British Grasses; an Introduction to the Study of the Gramineæ of Great Britain and Ireland. By M. PLUES. Crown 8vo, with 16 Coloured Plates and 100 Wood Engravings, 10*s.* 6*d.*

Familiar Indian Flowers. By LENA LOWIS. 4to, 30 Coloured Plates, 31*s.* 6*d.*

Botanical Names for English Readers. By RANDAL H. ALCOCK. 8vo, 6*s.*

Elementary Lessons in Botanical Geography. By J. G. BAKER, F.L.S. 3*s.*

FERNS.

British Ferns; an Introduction to the Study of the FERNS, LYCOPODS, and EQUISETA indigenous to the British Isles. With Chapters on the Structure, Propagation, Cultivation, Diseases, Uses, Preservation, and Distribution of Ferns. By M. PLUES. Crown 8vo, with 16 Coloured Plates, and 55 Wood Engravings, 10*s*. 6*d*.

The British Ferns; Coloured Figures and Descriptions, with Analysis of the Fructification and Venation of the Ferns of Great Britain and Ireland. By Sir W. J. HOOKER, F.R.S. Royal 8vo, 66 Coloured Plates, £2 2*s*.

Garden Ferns; Coloured Figures and Descriptions with Analysis of the Fructification and Venation of a Selection of Exotic Ferns, adapted for Cultivation in the Garden, Hothouse, and Conservatory. By Sir W. J. HOOKER, F.R.S. Royal 8vo, 64 Coloured Plates, £2 2*s*.

Filices Exoticæ; Coloured Figures and Description of Exotic Ferns. By Sir W. J. HOOKER, F.R.S. Royal 4to, 100 Coloured Plates, £6 11*s*.

Ferny Combes; a Ramble after Ferns in the Glens and Valleys of Devonshire. By CHARLOTTE CHANTER. Third Edition. Fcap. 8vo, 8 Coloured Plates and a Map of the County, 5*s*.

MOSSES.

Handbook of British Mosses, containing all that are known to be natives of the British Isles. By the Rev. M. J. BERKELEY, M.A., F.L.S. 24 Coloured Plates, 21*s*.

Synopsis of British Mosses, containing Descriptions of all the Genera and Species (with localities of the rarer ones) found in Great Britain and Ireland. By CHARLES P. HOBKIRK, President of the Huddersfield Naturalist's Society. Crown 8vo, 7*s*. 6*d*.

SEAWEEDS.

British Seaweeds; an Introduction to the Study of the Marine ALGÆ of Great Britain, Ireland, and the Channel Islands. By S. O. GRAY. Crown 8vo, with 16 Coloured Plates, 10*s*. 6*d*.

Phycologia Britannica; or, History of British Seaweeds. Containing Coloured Figures, Generic and Specific Characters, Synonyms and Descriptions of all the Species of Algæ inhabiting the Shores of the British Islands. By Dr. W. H. HARVEY, F.R.S. New Edition. Royal 8vo, 4 vols. 360 Coloured Plates, £7 10*s*.

Phycologia Australica; a History of Australian Seaweeds, comprising Coloured Figures and Descriptions of the more characteristic Marine Algæ of New South Wales, Victoria, Tasmania, South Australia, and Western Australia, and a Synopsis of all known Australian Algæ. By Dr. W. H. HARVEY, F.R.S. Royal 8vo, Five Vols., 300 Coloured Plates, £7 13*s*.

FUNGI.

Outlines of British Fungology, containing Characters of above a Thousand Species of Fungi, and a Complete List of all that have been described as Natives of the British Isles. By the Rev. M. J. BERKELEY, M.A., F.L.S. 24 Coloured Plates, 30*s*.

The Esculent Funguses of England. Containing an Account of their Classical History, Uses, Characters, Development, Structure, Nutritious Properties, Modes of Cooking and Preserving, &c. By C. D. BADHAM, M.D. Second Edition. Edited by F. CURREY, F.R.S. 12 Coloured Plates, 12*s*.

Clavis Agaricinorum; an Analytical Key to the British Agaricini, with Characters of the Genera and Sub-genera. By WORTHINGTON G. SMITH, F.L.S. 6 Plates, 2*s*. 6*d*.

SHELLS AND MOLLUSKS.

Testacea Atlantica; or, the Land and Freshwater Shells of the Azores, Madeiras, Salvages, Canaries, Cape Verdes, and Saint Helena. By T. VERNON WOLLASTON, M.A., F.L.S. Demy 8vo, 25*s*.

Elements of Conchology; an Introduction to the Natural History of Shells, and of the Animals which form them. By LOVELL REEVE, F.L.S. Royal 8vo, Two Vols., 62 Coloured Plates, £2 16*s*.

Conchologia Iconica; or, Figures and Descriptions of the Shells of Mollusks, with remarks on their Affinities, Synonymy, and Geographical Distribution. By LOVELL REEVE, F.L.S., and G. B. SOWERBY, F.L.S., complete in Twenty Vols., 4to, with 2727 Coloured Plates, half-calf, £178.

A detailed list of Monographs and Volumes may be had.

Conchologia Indica; Illustrations of the Land and Freshwater Shells of British India. Edited by SYLVANUS HANLEY, F.L.S., and WILLIAM THEOBALD, of the Geological Survey of India. Complete in One Vol., 4to, with 160 Coloured Plates, £8 5*s*.

The Edible Mollusks of Great Britain and Ireland, with the Modes of Cooking them. By M. S. LOVELL. Crown 8vo, with 12 Coloured Plates, 8*s*. 6*d*.

INSECTS.

The Lepidoptera of Ceylon. By F. MOORE, F.L.S. Parts I. to VII. Medium 4to, each with 18 Plates, to Subscribers only, 31*s*. 6*d*., coloured; 16*s*., uncoloured. Also Vol. I. (Rhopalocera), complete, cloth, gilt top, £7 15*s*.; to Subscribers for the entire work, £6 10*s*. Published under the auspices of the Government of Ceylon.

The Butterflies of Europe; Illustrated and Described. By HENRY CHARLES LANG, M.D., F.L.S. Super-royal 8vo, Parts I. to XIII., each, with 4 Coloured Plates, 3*s*. 6*d*. To be completed in about 20 Parts.

The Larvæ of the British Lepidoptera, and their Food Plants. By OWEN S. WILSON. With Life-size Figures, drawn and coloured from Nature by ELEANORA WILSON. Super-royal 8vo. With 40 elaborately-coloured Plates, containing upwards of 600 figures of Larvæ, 63*s*.

British Insects. A Familiar Description of the Form, Structure, Habits, and Transformations of Insects. By E. F. STAVELEY, Author of "British Spiders." Crown 8vo, with 16 Coloured Plates and numerous Wood Engravings, 14*s*.

British Beetles; an Introduction to the Study of our indigenous COLEOPTERA. By E. C. RYE. Crown 8vo 16 Coloured Steel Plates, and 11 Wood Engravings, 10*s*. 6*d*.

British Bees; an Introduction to the Study of the Natural History and Economy of the Bees indigenous to the British Isles. By W. E. SHUCKARD. Crown 8vo, 16 Coloured Plates, and Woodcuts of Dissections, 10*s*. 6*d*.

British Butterflies and Moths; an Introduction to the Study of our Native LEPIDOPTERA. By H. T. STAINTON. Crown 8vo, 16 Coloured Plates, and Wood Engravings, 10*s*. 6*d*.

British Spiders; an Introduction to the Study of the ARANEIDÆ found in Great Britain and Ireland. By E. F. STAVELEY. Crown 8vo, 16 Coloured Plates, and 44 Wood Engravings, 10*s*. 6*d*.

Harvesting Ants and Trap-door Spiders; Notes and Observations on their Habits and Dwellings. By J. T. MOGGRIDGE, F.L.S. With a SUPPLEMENT of 160 pp. and 8 additional Plates, 17*s*. The Supplement separately, cloth, 7*s*. 6*d*.

Curtis's British Entomology. Illustrations and Descriptions of the Genera of Insects found in Great Britain and Ireland, Containing Coloured Figures, from Nature, of the most rare and beautiful Species, and in many instances, upon the plants on which they are found. Eight Vols., Royal 8vo, 770 Coloured Plates, £28.

Or in Separate Monographs.

Orders.	*Plates.*	£ *s.* *d.*	*Orders.*	*Plates.*	£ *s.* *d.*
Aphaniptera .	2	0 2 0	Hymenoptera . .	125	6 5 0
Coleoptera . . .	256	12 16 0	Lepidoptera . .	193	9 13 0
Dermaptera . . .	1	0 1 0	Neuroptera . . .	13	0 13 0
Dictyoptera . . .	1	0 1 0	Omaloptera . . .	6	0 6 0
Diptera	103	5 3 0	Orthoptera . . .	5	0 5 0
Hemiptera . . .	32	1 12 0	Strepsiptera . .	3	0 3 0
Homoptera . . .	21	1 1 0	Trichoptera . .	9	0 9 0

"Curtis's Entomology," which Cuvier pronounced to have "reached the ultimatum of perfection," is still the standard work on the Genera of British Insects. The Figures executed by the author himself, with wonderful minuteness and accuracy, have never been surpassed, even if equalled. The price at which the work was originally published was £43 16*s*.

Insecta Britannica; Vol. III., Diptera. By Francis Walker, F.L.S. 8vo, with 10 Plates, 25*s*.

ANTIQUARIAN.

Sacred Archæology; a Popular Dictionary of Ecclesiastical Art and Institutions from Primitive to Modern Times. Comprising Architecture, Music, Vestments, Furniture Arrangement, Offices, Customs, Ritual Symbolism, Ceremonial Traditions, Religious Orders, &c., of the Church Catholic in all ages. By Mackenzie E. C. Walcott, B.D. Oxon., F.S.A., Precentor and Prebendary of Chichester Cathedral. Demy 8vo, 18*s*.

A Manual of British Archæology. By Charles Boutell, M.A. 20 Coloured Plates, 10*s*. 6*d*.

The Antiquity of Man; an Examination of Sir Charles Lyell's recent Work. By S. R. Pattison, F.G.S. Second Edition. 8vo, 1*s*.

MISCELLANEOUS.

Handbook of the Vertebrate Fauna of Yorkshire; being a Catalogue of Mammals, Birds, Reptiles, Amphibians, and Fishes, which are or have been found in the County. By W. E. CLARKE and W. D. ROEBUCK. 8vo, 8*s*. 6*d*.

Report on the Forest Resources of Western Australia. By Baron FERD. MUELLER, C.M.G., M.D., Ph.D., F.R.S., Government Botanist of Victoria. Royal 4to, 20 Plates, 12*s*.

West Yorkshire; an Account of its Geology, Physical Geography, Climatology, and Botany. By J. W. DAVIS, F.L.S., and F. ARNOLD LEES, F.L.S. Second Edition, 8vo, 21 Plates, many Coloured, and 2 large Maps, 21*s*.

Handbook of the Freshwater Fishes of India; giving the Characteristic Peculiarities of all the Species at present known, and intended as a guide to Students and District Officers. By Capt. R. BEAVAN, F.R.G.S. Demy 8vo, 12 plates, 10*s*. 6*d*.

Natal; a History and Description of the Colony, including its Natural Features, Productions, Industrial Condition and Prospects. By HENRY BROOKS, for many years a resident. Edited by Dr. R. J. MANN, F.R.A.S., F.R.G.S., late Superintendent of Education in the Colony. Demy 8vo, with Maps, Coloured Plates, and Photographic Views, 21*s*.

St. Helena. A Physical, Historical, and Topographical Description of the Island, including its Geology, Fauna, Flora, and Meteorology. By J. C. MELLISS, A.I.C.E., F.G.S., F.L.S. In one large Vol., Super-royal 8vo, with 56 Plates and Maps, mostly coloured, 42*s*.

Lahore to Yarkand. Incidents of the Route and Natural History of the Countries traversed by the Expedition of

1870, under T. D. Forsyth, Esq., C.B. By George Henderson, M.D., F.L.S., F.R.G.S., and Allan O. Hume, Esq., C.B., F.Z.S. With 32 Coloured Plates of Birds, 6 of Plants, 26 Photographic Views, Map, and Geological Sections, 42*s*.

The Birds of Sherwood Forest; with Observations on their Nesting, Habits, and Migrations. By W. J. Sterland. Crown 8vo, 4 plates. 7*s*. 6*d*., coloured.

The Young Collector's Handy Book of Recreative Science. By the Rev. H. P. Dunster, M.A. Cuts, 3*s*. 6*d*.

A Survey of the Early Geography of Western Europe, as connected with the First Inhabitants of Britain, their Origin, Language, Religious Rites, and Edifices. By Henry Lawes Long, Esq. 8vo, 6*s*.

The Geologist. A Magazine of Geology, Palæontology, and Mineralogy. Illustrated with highly-finished Wood Engravings. Edited by S. J. Mackie, F.G.S., F.S.A. Vols. V. and VI., each with numerous Wood Engravings, 18*s*. Vol. VII., 9*s*.

Everybody's Weather-Guide. The use of Meteorological Instruments clearly explained, with directions for securing at any time a probable Prognostic of the Weather. By A. Steinmetz, Esq., Author of "Sunshine and Showers," &c. 1*s*.

The Artificial Production of Fish. By Piscarius. Third Edition. 1*s*.

The Gladiolus: its History, Cultivation, and Exhibition. By the Rev. H. Honywood Dombrain, B.A. 1*s*.

Meteors, Aerolites, and Falling Stars. By Dr. T. L. Phipson, F.C.S. Crown 8vo, 25 Woodcuts and Lithographic Frontispiece, 6*s*.

The Zoology of the Voyage of H.M.S. *Samarang*, under the command of Captain Sir Edward Belcher, C.B., during the Years 1843-46. By Professor OWEN, Dr. J. E. GRAY, Sir J. RICHARDSON, A. ADAMS, L. REEVE, and A. WHITE. Edited by ARTHUR ADAMS, F.L.S. Royal 4to, 55 Plates, mostly coloured, £3 10*s*.

Papers for the People. By ONE OF THEM. No. 1, Our Land. 8vo, 6*d*. (By Post, 7*d*. in stamps.)

The Royal Academy Album; a Series of Photographs from Works of Art in the Exhibition of the Royal Academy of Arts, 1875. Atlas 4to, with 32 fine Photographs, cloth, gilt edges, £6 6*s*.; half-morocco, £7 7*s*.

The same for 1876, with 48 beautiful Photo-prints, cloth, £6 6*s*.; half-morocco, £7 7*s*. Small Edition, Royal 4to, cloth, gilt edges, 63*s*.

On Intelligence. By H. TAINE, D.C.L. Oxon. Translated from the French by T. D. HAYE, and revised, with additions, by the Author. Complete in One Vol., 18*s*.

Manual of Chemical Analysis, Qualitative and Quantitative; for the use of Students. By Dr. HENRY M. NOAD, F.R.S. New Edition. Crown 8vo, 109 Wood Engravings, 16*s*. Or, separately, Part I., "QUALITATIVE," New Edition, new Notation, 6*s*.; Part II., "QUANTITATIVE," 10*s*. 6*d*.

Live Coals; or, Faces from the Fire. By L. M. BUDGEN, "Acheta," Author of "Episodes of Insect Life," &c. Dedicated, by Special Permission, to H.R.H. Field Marshal the Duke of Cambridge. Royal 4to, 35 Original Sketches printed in colours, 21*s*.

Caliphs and Sultans; being tales omitted in the ordinary English version of "The Arabian Nights' Entertainments," freely rewritten and rearranged. By S. HANLEY, F.L.S. 6*s*.

PLATES.

Floral Plates, from the Floral Magazine. Beautifully Coloured, for Screens, Scrap-books, Studies in Flower-painting, &c. 6*d.* and 1*s.* each. Lists of over 700 varieties, One Stamp.

Botanical Plates, from the Botanical Magazine. Beautifully-coloured Figures of new and rare Plants. 6*d.* and 1*s.* each. Lists of over 2000, Three Stamps.

SERIALS.

The Botanical Magazine. Figures and Descriptions of New and rare Plants. By Sir J. D. HOOKER, C.B., F.R.S. Monthly, with 6 Coloured Plates, 3*s.* 6*d.* Annual subscription, post free, 42*s.*

Re-issue of the Third Series, in Monthly Vols., 42*s.* each; to Subscribers for the entire Series, 36*s.* each.

The Lepidoptera of Ceylon. By F. MOORE. 16*s.*, plain; 31*s.* 6*d.*, coloured.

The Butterflies of Europe. By Dr. LANG, F.L.S. Monthly, with 4 Coloured Plates, 3*s.* 6*d.* Subscription for the complete Work (20 Parts), in advance, 60*s.*

FORTHCOMING WORKS.

Flora of India. By Sir J. D. HOOKER and others. Part XI.

Natural History of Plants. By Prof. BAILLON.

Flora of Tropical Africa. By Prof. OLIVER.

Flora Capensis. By Prof. DYER.

London:

L. REEVE & CO., 5, HENRIETTA STREET, COVENT GARDEN.

PRINTED BY GILBERT AND RIVINGTON, LIMITED, ST. JOHN'S SQUARE, LONDON.